Magnetocaloric Effect in Perovskite Manganites

H. Gencer[a], V.S. Kolat[b], T. Izgi[c], N. Bayri[d], S. Atalay[e]

Inonu University, Science and Arts Faculty, Physics Department, Malatya 44280, Turkey

[a]huseyin.gencer@inonu.edu.tr, [b]kolat.veli@inonu.edu.tr,
[c]tekin.izgi@inoun.edu.tr,[d]nevzat.bayri@inonu.edu.tr, [e]selcuk.atalay@inonu.edu.tr

Edited by

Rajshree B. Jotania[1], Sami H. Mahmood[2]

[1]Department of Physics, Electronics and Space science, University school of sciences, Gujarat University, Ahmedabad 380 009, India

[2]Department of Physics, The University of Jordan,Amman-11942, Jordan

Published by **Materials Research Forum LLC**
Millersville, PA 17551, USA

Published as part of the book series
Materials Research Foundations
Volume 81 (2020)
ISSN 2471-8890 (Print)
ISSN 2471-8904 (Online)

Print ISBN 978-1-64490-92-5
ePDF ISBN 978-1-64490-093-2

Distributed worldwide by

Materials Research Forum LLC
105 Springdale Lane
Millersville, PA 17551
USA
http://www.mrforum.com

Printed in the United States of America
10 9 8 7 6 5 4 3 2 1

Table of Contents

1. Introduction

In today's modern society, cooling at room temperature has become an indispensable technology at every point of daily life, such as houses, public buildings, and air conditioning in vehicles, to store food in homes and markets. So far, refrigerators based on gas compression and expansion logic have been widely used for cooling applications. The use of chlorofluorocarbons (CFC) and hydrochlorofluorocarbons (HCFC) gases as a refrigerant in conventional cooling technology has brought with it serious environmental concerns due to global warming, especially damaging the ozone layer [1]. According to Montreal Protocol, the prohibition of the use of CFC and HCFC gases, replacing them with hydrofluorocarbons (HFC) that do not contain chlorine and therefore do not harm the ozone layer, but it does not solve the problem. This is because HFC is a greenhouse gas with global warming potential higher than CO_2 [2]. According to the Montreal Protocol, the use of HFC gases will be banned in the following years [3]. Therefore, due to severe environmental concerns, alternative technologies should be introduced that can offer more appealing solutions to environmental problems rather than existing gas cooling technology.

In recent years, the development of new magnetic refrigeration (MR) technology based on the magnetocaloric effect (MCE) has brought a new and promising alternative to conventional gas cooling technology [4]. In a magnetic refrigerator based on the Gd element, although the cooling efficiency reaches 60% of the theoretical limit, this ratio remains only 40% in gas-compressed refrigerators [5]. At the same time, magnetic refrigeration is an environmentally friendly and cost-effective technology that saves up to 30% energy compared to conventional gas refrigeration technology. The use of magnetic refrigeration technology with high energy efficiency will reduce CO_2 emissions by reducing fossil fuel consumption. It also prevents the use of gases that damage the ozone layer (CFC), greenhouse gases and hazardous chemicals. Magnetic refrigeration has significant advantages such as small volume, chemical stability, low cost, non-toxic and not causing sound pollution. Magnetic refrigeration has been used in scientific applications for a long time for refrigeration under 1 K [6]. However, since most ferromagnetic materials do not exhibit adequate magnetocaloric properties around room temperature, there are no commercial practices of magnetic refrigeration at room temperature. For this reason, a large portion of the studies in the field of magnetic refrigeration is still related to the discovery of materials at various temperatures (especially at room temperature) and with sizeable magnetocaloric effect. So far, many promising materials have been reported to be candidates for magnetic refrigeration, such as Ni-Mn-Ga compounds [7], Mn-As-Sb compounds [8], La-Fe-Co-Si compounds [9-12], Mn-Fe-P-As compounds [13], and La-Ca-Sr-Mn−O manganites [14-36]. Among these magnetocaloric materials mentioned, perovskite manganites have been prominent because of its significant properties, such as extremely large magnetic entropy and adiabatic temperature variations, much smaller thermal or magnetic hysteresis, higher chemical stability for low cost and long-term use. Also, Curie temperature and saturation magnetization can be tailored by changing doping element and doping concentration, this makes manganites a suitable material for magnetic refrigeration at different temperatures.

2. Magnetocaloric effect

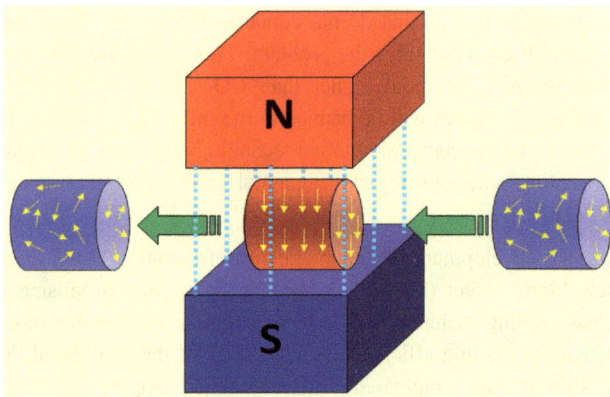

Fig. 1. Schematic representation of the magnetocaloric effect.

In 1881, the magnetocaloric effect (MCE), first observed by German scientist E. Warburg [37] on a piece of iron, refers to the change occurring at its temperature by applying a magnetic field to a material. This substantial truth is directly related to the entropy of the material. When a magnetic field is applied to a material isolated from its environment, randomly oriented magnetic moments of the material are forced to align in the same direction. Hence, the magnetic entropy of the material is reduced. As a result, the system increases its temperature by a few degrees to compensate the decreasing entropy balance. Therefore, the material refrigerates the environment by making heat absorption. This situation underlies magnetic refrigeration technology [38]. When the magnetic field is removed, the magnetic moments gain a random oriented direction again, the entropy of the system increases and the temperature of material decreases (Fig. 1).

2.1 Historical development of magnetic refrigeration

The first studies on the origin of the magnetocaloric effect were initiated by W.F. Giauque [6] in 1927 and it has been shown that extremely low temperatures (0.25 K) can be reached by using paramagnetic salts. With this study, experimental studies aimed to reducing the low temperatures (mK) to ultra-low temperatures (µK) have accelerated. Today, this technology is used to reach extremely low temperatures.

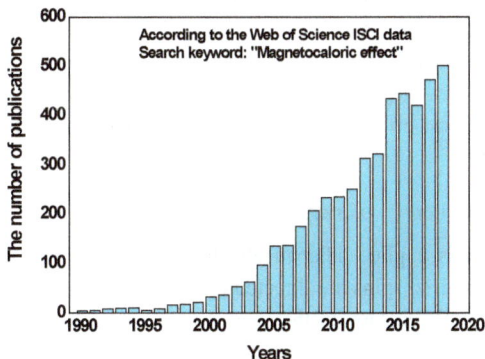

Fig. 2. According to Web of Science and ISCI (International Science Citation Index) data, distribution of the number of publications associated with magnetocaloric effect by years.

In the past decades and today, the studies on the magnetocaloric effect and its technological applications are intensively going on. In particular, studies that encourage the idea of developing magnetocaloric refrigeration that can operate at room temperature and can be an alternative to conventional refrigerations have been provided with the studies made with gadolinium and Gd-based compounds. Studies on the magnetocaloric properties of Gd and Gd-based compounds, which were introduced by G.V. Brown [39] in 1976, have accelerated the development of modern magnetic refrigerants that can operate around room temperature. First magnetocaloric effect studies in LaMnO-based, doped perovskite-type film compounds, were conducted by D.T. Morelli [40] in 1996. In studies with LaAMnO (A = Ca, Ba, Sr) film compounds, positive results were obtained regarding the magnetocaloric effect. In parallel to this study, in 1996, the higher magnetocaloric effect was observed by X. X. Zhang [41] in LaCaMnO ceramic bulk materials. Thus, studies carried out with LaMnO-based, doped perovskite-type compounds have gained speed. The high magnetic entropy change, called GMCE (Giant Magnetocaloric Effect), was first observed in 1997 by V.K. Pecharsky and K.A. Jr. Gschneidner [42], in the compound of $Gd_5(Si,Ge)_4$, and was several times larger than pure Gd. In addition, we can summarize some of the different compounds that are still being studied today and exhibit GMCE, such as GdDy, GdTy, Gd(Si-Ge), La(Fe-Si)H, MnFe (P-As), and especially in the last few years, FeSiB-based amorphous and doped amorphous compounds.

Fig. 2 shows the distribution of the number of publications by year related to magnetocaloric effect published in international journals which belong to ISCI (International Science Citation Index) taken from Web of Science since 1994. As a result, as can be seen from the bar graph, the scientific studies on the development of materials that have magnetocaloric effect and superior magnetocaloric effect are showing a rapid increase until today.

2.2   Basic thermodynamics of magnetocaloric effect

If the magnetic field is applied adiabatically to a ferromagnetic material around Curie temperature (T_C), unpaired spins are directed towards the direction of the magnetic field. As a result, the magnetic entropy of the solid matter decreases and the lattice entropy of the sample increases. Due to the increase in lattice entropy, the sample increases its heat to allow the decrease in magnetic entropy to rise again. As a result, when the field is eliminated, the spin turns back randomly, the magnetic entropy increases, and the heat of the sample decreases with the lattice entropy.

Under constant pressure, the entropy of a magnetic solid, S(T, H), can be written as three different entropy sums [5-22,43].

$$S(T,H) = S_M(T,H) + S_{Lat}(T) + S_{El}(T) \tag{1}$$

Here, S_M, magnetic, S_{Lat}, lattice and S_{El}, represent electronic entropy. The change of magnetic entropy of a magnetic material is given by:

$$\Delta S_M(T)_{\Delta H} = \left[S(T)_{H_1} - S(T)_{H_0} \right]_T \tag{2}$$

Similarly, the adiabatic temperature variation of a magnetic material is given by:

$$\Delta T_{ad}(T)_{\Delta H} = \left[T(S)_{H_1} - T(S)_{H_0} \right]_S \tag{3}$$

The term adiabatic temperature change (ΔT_{ad}) and isothermal magnetic entropy change (ΔS_M) relate to the terms magnetization, magnetic field strength and heat capacity under constant pressure and constant temperature. According to Maxwell's Equations [43];

$$\left(\frac{\partial S(T,H)}{\partial H}\right)_T = \left(\frac{\partial M(T,H)}{\partial T}\right)_H \tag{4}$$

a correlation can be established. With the integration of this equation;

$$\Delta S_M(T)_{\Delta H} = \int_{H_0}^{H_1} dS_M(T,H)_T = \int_{H_0}^{H_1} \left(\frac{\partial M(T,H)}{\partial T}\right)_H dH \tag{5}$$

and

$$\Delta T_{ad}(T)_{\Delta H} = \int_{H_0}^{H_1} dT(T,H) = -\int_{H_0}^{H_1} \left(\frac{T}{C(T,H)}\right)_H \left(\frac{\partial M(T,H)}{\partial T}\right)_H dH \tag{6}$$

Both $\Delta S_M(T)_{\Delta H}$ and $\Delta T_{ad}(T)_{\Delta H}$ depend on temperature and ΔH. The characteristic behavior of both magnetocaloric effects depends on the properties of the material. Therefore, without experimental measurements, knowledge about these behaviors is complicated to predict.

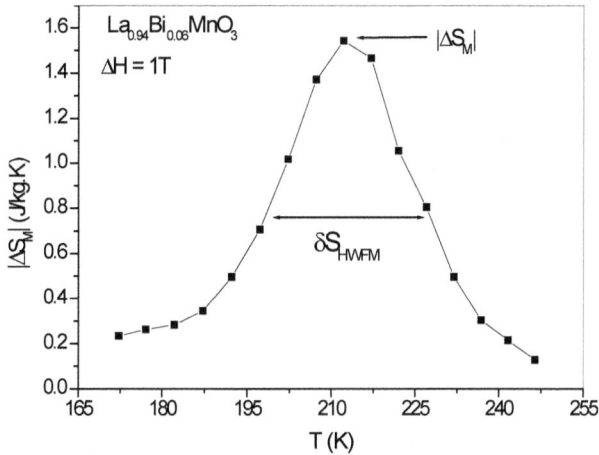

Fig. 3. Calculating the RCP value from entropy change for the $La_{0.97}Bi_{0.06}MnO_3$ compound [30].

As is known, the magnetocaloric effect is the physical basis of magnetic refrigeration systems. One of the technologically essential parameters such as maximum magnetic entropy change and operating temperature is the relative cooling power (RCP), which represents the magnetic cooling efficiency. In short, this parameter is expressed in term of the product of the maximum value of the $|\Delta S_M|$ or ΔT_{ad} and the half-height temperature width (δT_{FWHM}) (Fig. 3).

$$RCP(S) = \left| \Delta S_M^{max} \right| \times \delta T_{FWHM} \tag{7}$$

$$RCP(T) = \Delta T_{ad} \times \delta T_{FWHM} \tag{8}$$

2.3   Measurement of magnetocaloric effect

We can assemble the measurement methods of magnetocaloric effect in two main groups. First, the magnetocaloric effect can be measured using direct methods [43]. Secondly, it can be calculated using indirect methods like using magnetization or heat capacity measurements [18,19,43]. Whether direct or indirect methods are used, the result of measurements or calculations is a function of the temperature and magnetic field. In comparison, both techniques have some advantages and some disadvantages.

Direct measurement methods only give adiabatic temperature change (ΔT_{ad}). Temperature values are found without any processing of data, and the magnetocaloric effect is easily obtained by taking the difference between the two temperature values. However, direct measurement usually has time delays, and it is difficult to measure for small changes in temperature. In a direct measurement, massive experimental mistakes are unavoidable if the measuring devices are not calibrated well or the material is not properly isolated.

Indirect MCE measurements allow the calculation of both $\Delta T_{ad}(T)_{\Delta H}$ and $\Delta S_M(T)_{\Delta H}$ using experimental heat capacity data or will enable the calculation of $\Delta S_M(T)_{\Delta H}$ alone using experimental magnetization measurements. Indirect measurement gives practical results in any temperature range. However, experimental data must be processed to calculate MCE.

2.3.1   Direct measurements

2.3.1.1 Measurements under variable magnetic field

Thermal isolation of the sample is of great significance in the direct measurement method. In this method, the initial temperature of a thermally isolated sample is measured in an initial field ($T_i(H_i)$). Then, the area is subtracted from the starting value (H_i) to the final value (H_f), and the final temperature of the sample is measured ($T_f(H_f)$). The difference between the temperature values obtained by using these two field values. Thus, the adiabatic temperature change can be expressed as:

$$\Delta T_{ad}(T_i)_{\Delta H} = T_f - T_i \tag{9}$$

The form of the magnetic field applied to the sample is as pulsed when the field applied or removed. On the other hand, it can be in the form of steps with a magnetic field change of ~10 kOe/s [43]. Weiss and Forrer first proposed the direct measurement method in 1926 by using an electromagnet to create and eliminate the field using the switch-on technique [44]. Later, in 1969, Clark and Callen made the first measurements using the yttrium iron core under a powerful magnetic field (above 110 kOe) [45]. In both studies, a thermocouple was used to measure the temperature. In 1988, Green used the same method but used a superconducting selenoid instead of an electromagnet to reach higher fields [43]. Another method of direct measurement is the differential thermocouple method proposed by Kuhrt in 1985, which provides more accurate results [43].

2.3.1.2 Measurements under static magnetic field

The field generated by an electromagnet is approximately 20 kOe. However, if a superconducted solenoid is used, this field value can be increased to 100 kOe or above. A heat distribution occurs due to the magnetocaloric behavior of the sample to attain the desired field value of electromagnets in the elapsed time. In order to eliminate this undesirable situation, the observations made by Tishin in 1988 [46] have showed that the time to reach the desired value of the field for temperatures above 30 K, it should not be greater than 10 seconds. In order to overcome these difficulties, which are related to the duration of the applied field to reach the desired value, the static magnetic field method based on the logic of fast positioning of the sample into the static magnetic field of a superconducting solenoid was first developed by Nikitin [47] in 1985, Gopal [48] in 1987 and Tishin [46] in 1988. According to this method, the sample is initially outside the solenoid, and when solenoid reaches the desired field, the sample is quickly placed at the center of the solenoid (~1s), and the temperature is measured.

2.3.2 Indirect measurements

2.3.2.1 Magnetization measurements

Magnetic entropy change (ΔS_M) using experimental isothermal magnetization data (M(H)) can be calculated using Eq.(5). In 1993, McMichael [49] introduced the following simple formula for numerical computation of ΔS_M.

$$\left| \Delta S_M \right| = \sum_i \frac{1}{T_{i+1} - T_i} \left(M_i - M_{i+1} \right) \Delta H_i \tag{10}$$

Fig. 4 displays the calculation of magnetic entropy change (ΔS_M) using typical M-H curves for LaCaMnO structure. Accordingly, the area between the M-H curves in the two different temperature ranges (T and $T+\Delta T$) can be calculated and the ΔS_M change corresponding to the temperature range ($T+\Delta T/2$) can be found according to Eq. (11) as given below.

$$\Delta S_M \left(T + \frac{\Delta T}{2}, H \right) \approx -\frac{1}{\Delta T} \times \boxed{\text{Area}} \tag{11}$$

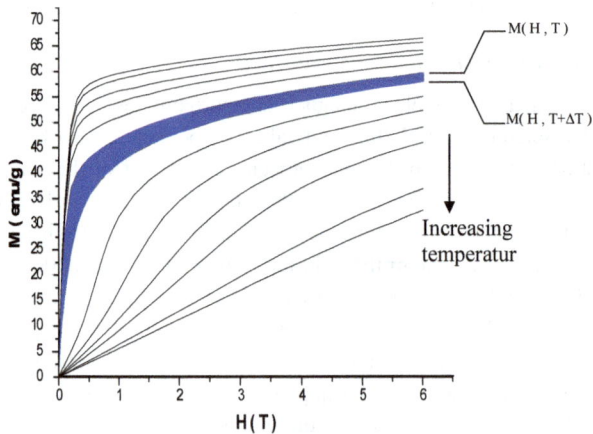

Fig. 4. Calculation of magnetic entropy change (ΔS_M) from the field between M-H curves.

2.3.2.2 Specific heat measurements

MCE and magnetic entropy change can be determined from the measurements of temperature dependence of heat capacity in different magnetic fields. This method was developed by Brown [39] in 1976 and Gschneidner [50] in 1996. Entropy change of material under magnetic field change $\Delta H = (H_2\text{-}H_1)$ can be calculated as follows,

$$\Delta S_M (T)_{\Delta H} = \left[\int_0^T \frac{C(T)_{H_2}}{T} dT - \int_0^T \frac{C(T)_{H_1}}{T} dT \right] \tag{12}$$

2.3.3 Semi theoretical determination methods

2.3.3.1 Determination from resistivity measurements

Many researchers have shown that there is a strong correlation between electrical and magnetic properties. [51,52]. In manganites, both the magnetocaloric effect and magnetoresistance properties usually occur around the magnetic phase transition temperature. This shows that there is a correlation between magnetic entropy change and resistivity (ρ). In this context, Xiong [53] proposed a correlation between ΔS_M and ρ as:

$$\Delta S_M = -\alpha \int_0^H \left[\frac{\delta \ln(\rho)}{\delta T} \right]_H dH \tag{13}$$

Here α, is the adjustment parameter and reflects the magnetic properties of the material. The constant α is calculated using different functions ($\rho = \rho_0 \exp(-M^2/\alpha)$ [51], $\rho = \rho_0 \exp(-M^2/\alpha T)$ [52], $\rho = \rho_0 \exp(-M/\alpha)$ [53]) that correlate M and ρ. For $La_{0.67}Ca_{0.33}MnO_3$ compound, $\alpha = 21.74$ emu/g has been found [53]. Fig. 5 shows the magnetic entropy change obtained from conventional magnetization curves and resistance measurements for $La_{0.67}Ca_{0.33}MnO_3$ compound.

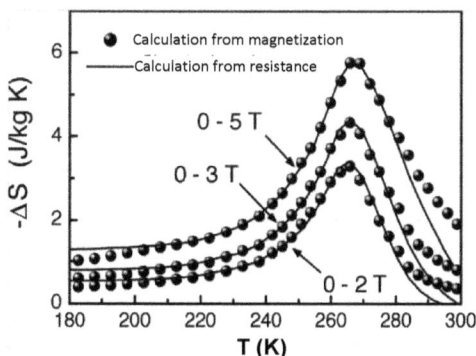

Fig. 5. Comparison of magnetic entropy change calculated from resistance measurements and magnetization curves [53].

2.3.3.2 Determination from landau theory

Landau expansion of magnetic free energy, F(M,T), in terms of magnetization,

$$F(M,T) = \frac{c_1(T)}{2} M^2 + \frac{c_3(T)}{4} M^4 + \frac{c_5(T)}{6} M^6 - MH \tag{14}$$

Here $c_1(T)$, $c_3(T)$ and $c_5(T)$ are called Landau coefficients, which depend on the temperature. From the condution of equilibrium $\partial F(M,T)/\partial M = 0$, magnetic entropy change can be obtained as:

$$\Delta S_m(T,H) = -\frac{1}{2}\frac{\partial c_1}{\partial T} M^2 - \frac{1}{4}\frac{\partial c_3}{\partial T} M^4 - \frac{1}{6}\frac{\partial c_5}{\partial T} M^6 \tag{15}$$

$c_1(T)$, $c_3(T)$ and $c_5(T)$ coefficients are found as the function of temperature using experimental magnetization curves (Fig. 6). Magnetic entropy change is determined using the obtained coefficients and Eq. (15).

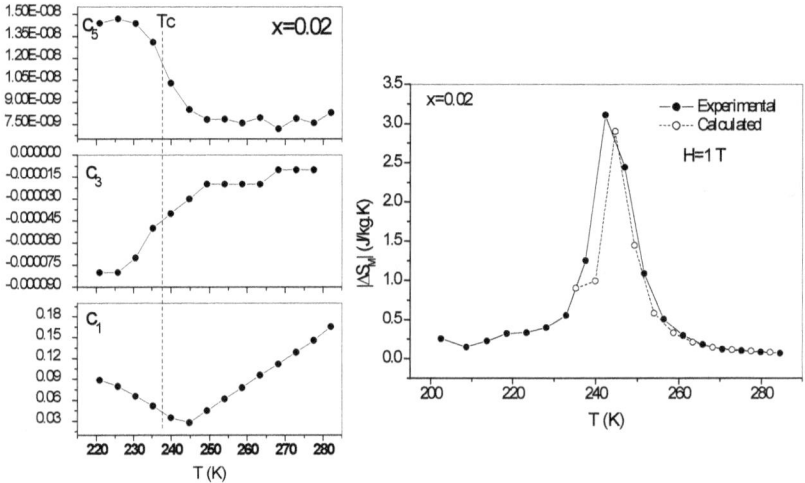

Fig. 6. $c_1(T.g/emu)$, $c_3(T.g^3/emu^3)$ and $c_5(T.g^5/emu^5)$ coefficients and calculated magnetic entropy change for $La_{0.67}Ca_{0.31}Mg_{0.02}MnO_3$ compound [33].

2.3.3.3 Determination from mean-field method

As it is known, the Landau theory relies on the logic of the expansion of magnetism in terms of its forces; it does not take into account the magnetic behavior in higher fields. Since this limitation is particularly significant in low temperatures and high fields, a generalized mean-field analysis method has been developed, which is also profoundly successful in determining the magnetocaloric properties of materials [54]. The general mean-field principle is used as defined in the form,

$$M(T,H) = f\left(\frac{H + H_{exc}}{T}\right) \tag{16}$$

to obtain the mean-field parameters from experimental magnetization data [55]. Here H_{exc} = λM is the exchange field and f is called the state function. Inverse f^{-1} (M) function is defined as:

$$\frac{H}{T} = f^{-1}(M) - \frac{H_{exc}}{T} \tag{17}$$

From experimental *M-H* curves, *H* and *T* values for specific magnetization values are determined, (Fig.7a), and graphs of *H/T* against *1/T* are drawn (Fig.7b). For each *M* value, the slope of the drawn curves is equal to H_{exc} ($= \lambda_1 M + \lambda_3 M^3 + ...$).

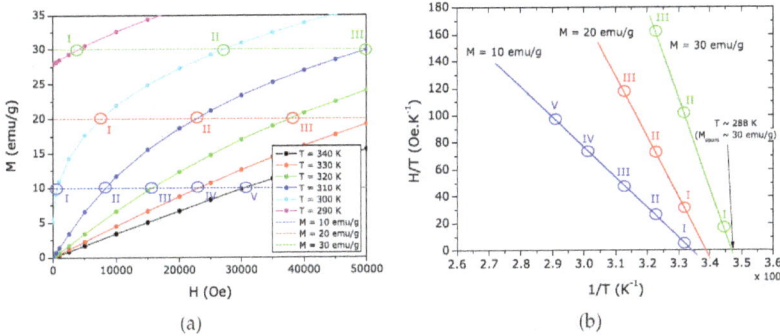

Fig.7(a). Obtaining mean-field data for specific magnetization values (M = 10, 20 and 30 emu/g) in M-H curves; Fig.7 (b). H/T-1/T curves [56].

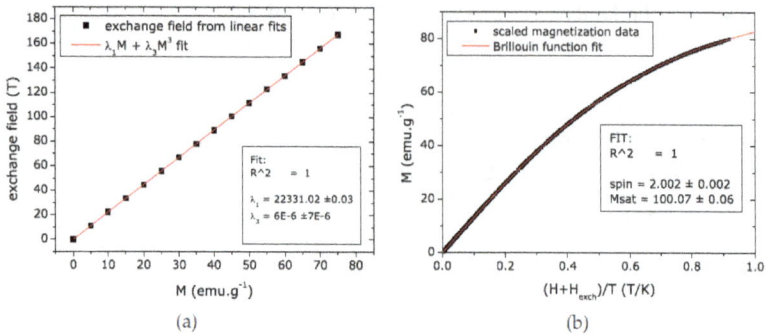

(a)   (b)

Fig. 8(a). Exchange field transformation in response to magnetism; Fig.8 (b). Transformation of M corresponding $(H + H_{exch})/T$ [56].

After the exchange field is determined, the graph is drawn in the second step, corresponding to the magnetization values of the exchange field values (Fig. 8a). After the exchange field is obtained, the second step of this method is to create a scale graph (f function), M against $(H + H_{exch})/T$, where the data defining the system is collected on a curve (Fig.8b).

The f scale function is related to the magnetic entropy ($f^{-1}(M) = \partial S_M / \partial M$), thus, the magnetic entropy change between the H_1 and H_2 fields

$$\Delta S_M = \int_{M/H_1}^{M/H_2} f^{-1} dM \qquad (18)$$

Magnetic entropy change is obtained by using the f function given in Eq. (18) and Fig. 8b. Fig.9 shows the magnetic entropy value obtained for $La_{0.665}Er_{0.035}Sr_{0.3}MnO_3$ compound [56].

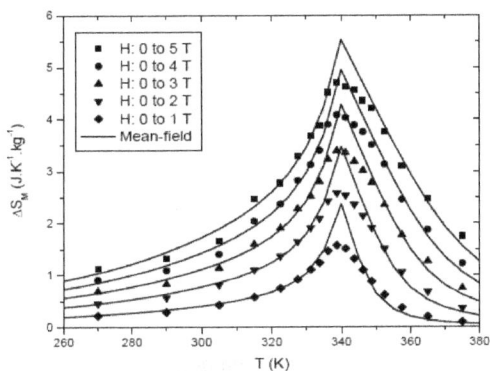

Fig. 9. Comparison of magnetic entropy change in La$_{0.665}$Er$_{0.035}$Sr$_{0.3}$MnO$_3$ compound with the Mean-field method and experimentally obtained [56].

2.3.3.4 Determination from a phenomenological model

In addition to experimental studies, magnetic and magnetocaloric properties of magnetic materials are studied using theoretical models [57]. From Hamad's theoretical model and magnetization measurements $(M(T))$, it is possible to determine the magnetocaloric properties of magnetic materials. In the studies carried out, it is claimed that this model is more accessible and gives more precise results than previously used theoretical models [58]. In this model, magnetization changes according to temperature

$$M(T) = \left(\frac{M_i - M_f}{2} \right) \tanh\left[A\, (T_c - T) \right] + B\,T + C$$

(19)

Here, M_i and M_f show the initial and final values of the ferromagnetic-paramagnetic transition region, as shown in Fig.10.

Here, $A = 2(B - S_c)/(M_i - M_f)$ and $C = (M_i + M_f)/2 - BT_c$.

Here, B gives the magnetization sensitivity $\left(= \dfrac{dM}{dT}\Big|_{T=T_i} \right)$ in a ferromagnetic state before

the transition. The S_c is magnetization sensitivity $\left(= \dfrac{dM}{dT}\Big|_{T=T_c} \right)$ at Curie temperature.
Magnetic entropy change is then obtained using Eq.(19) and Eq. (5).

$$\Delta S_M = \left(- A\, \frac{M_i - M_f}{2}\, \sec h^2 \left[A(T_c - T)\right] + B \right) H_{\max} \tag{20}$$

Similarly, adiabatic temperature change is given by:

$$\Delta T_{ad} = \frac{A\,T\,(M_i - M_f)}{2\,C_p} \left[\sec h^2 (A\,(T_c - T)) + B\right] H_{\max} \tag{21}$$

Here, C_p is the heat capacity per mol, in the constant magnetic field. In Fig.11, ΔS_M and ΔT_{ad} changes calculated in the 0.1 T magnetic field for $La_{0.94}Bi_{0.06}Mn_{1-x}Cr_xO_3$ ($x = 0$, 0.05, 0.1, 0.15, 0.2, 0.25) compounds are observed using the Eqs. 20, 21 and experimentally measured magnetization curves.

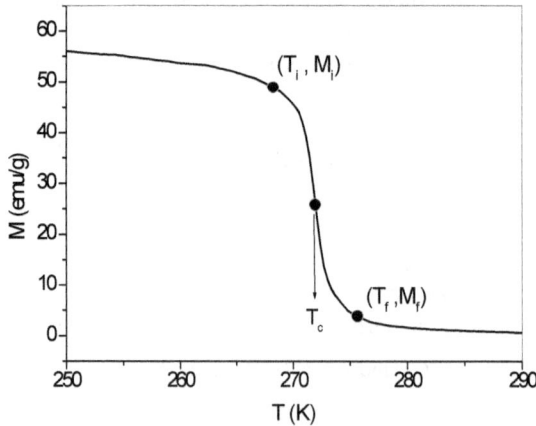

Fig.10. Change of magnetization for $La_{0.67}Ca_{0.33}MnO_3$ with temperature.

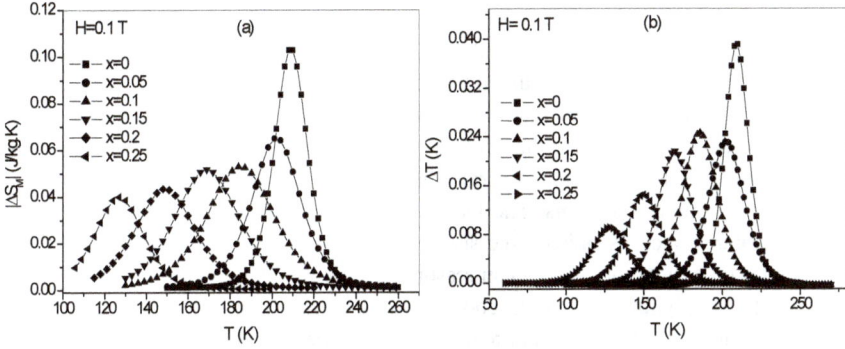

Fig.11. Magnetic entropy change and and adiabatic temperature change calculated in 0.1 T magnetic field for $La_{0.94}Bi_{0.06}Mn_{1-x}Cr_xO_3$ compound.

2.4   Magnetic cooling

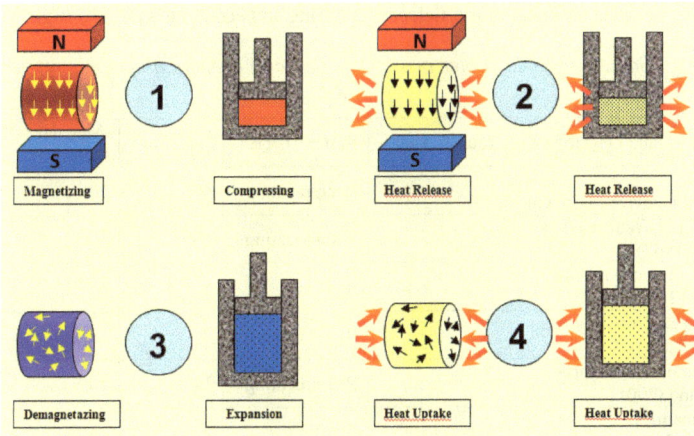

Fig.12. Schematic representation of magnetic cooling system and gas cycled commercial cooling system comparatively.

Since the discovery of the magnetocaloric effect, studies have been in progress on the development of magnetic cooling systems. The idea of magnetic cooling which has not yet been fully utilized as a commercial cooling system, has been used for more than 50 years in order to further cool the ultra-cold environments. Since magnetic cooling technology is capable of competing with today's commercial cooling systems, it is one of the greatest efforts to produce the cooling technology for the future. Also, since solid materials are used as refrigerant material, they have significant advantages compared to gas-cycled commercial systems. The fact that magnetic refrigerations do not carry any adverse features such as noise, excessive vibration, oil or gas leakage, abrasion, and dependence on gravitation, is the main reason for these materials to be seen as the future refrigerants. Simply, the cycle of the cooling process is shown in Fig.12 [23]. A magnetic refrigerant can be operated at room temperature, first developed by Brown [39] in 1976 and Steyert [59] in 1978, has increased interest in this field in recent years.

In Table 1, the development of magnetic cooling systems with room temperature operating characteristics are given by years. In recent years, the use of permanent magnets in prototype magnetic cooling systems is an exciting feature. Besides, systems developed in recent years have smaller volumes, and this is seen as an essential improvement for commercial use. When the historical development of magnetic cooling systems is examined, it is predicted that in a not-so-distant future, magnetic cooling systems will replace commercial refrigerants. The studies carried out today are continued under two central headings: the development of materials with magnetocaloric effect with superior efficiency, enduring and economic system design.

Table 1. Some magnetic cooler prototypes that can operate at room temperature [39,60].

Research and Development Group & Year	System Type	T_{span} (K)	Magnetic Field (kOe)
Brown The first device to use Gd (1976)	Reciprocating	47	70(S)
Ames Lab./Astronautics USA (1997)	Reciprocating	10	50(S)
Mater. Science Institute Spain (2000)	Rotating	5	9.5(P)
Chuba Electric/Toshiba Japan (2000)	Reciprocating	21	40(S)
University of Victoria England (2001)	Reciprocating	14	20(S)
Astronautics USA (2001)	Rotating	20	15(P)
Sichuan Inst. Tech./Nanjing Uni. China (2002)	Reciprocating	23	14(P)
Chuba Electric/Toshiba Japan (2002)	Reciprocating	27	6(P)
Chuba Electric/Toshiba Japan (2003)	Rotating	10	7.6(P)

Lab. d'Electrontechnique Grenoble France (2003)	Reciprocating	4	8(P)
The Universities of Victoria and Quebec, Canada (2004)	Reciprocating	47/51 14	20(P) 20(S)
Washington State University USA (2004)	Reciprocating	5	20(P)
Graduate School of Engineering of the Hokkaido University Japan (2005)	Reciprocating	10	20(P)
Chinese Academy of Science China (2006)	Reciprocating	18/10	15(P)
Baotou Research Institute of Rare Earth China (2006)	Reciprocating	5	15(P)
Astronautics Corporation of America USA (2007)	Rotating	12	15(P)
University of Victoria Canada (2007)	Rotating	13	14(P)
Sichuan University China (2007)	Rotating	11.5	15(P)
Riso National Laboratory Denmark (2007)	Reciprocating	9	12(P)
Bahl *et al.* (2008)	Reciprocating	6-7	14(E)
Hirano *et al.* (2009)	Reciprocating	2	23(P)
Cooltech Applications France (2009)	Reciprocating	16	11(P)
Campinas State University Brasil (2009)	Rotating	11	23(E)
Dupuis *et al.* (2009)	Reciprocating	7.8	8(P)
Korea Advanced Institute of Science and Technology Korea (2009)	Reciprocating	16	15.8(P)
University of Genoa Italy (2009)	Reciprocating	5	15,5(P)
Trevizoli *et al.* (2010)	Reciprocating	4.4	16.5(P)
Korea Advanced Institute of Science and Technology Korea (2011)	Reciprocating	14	15(P)
Balli *et al.* (2011)	Reciprocating	20	14.5(P)
Tura *et al.* (2011)	Rotating	29	14.7(P)
Technical University of Denmark Denmark (2011)	Reciprocating	5-10	10.3(P)
Park *et al.* (2012)	Reciprocating	26.8	14(P)
University of Victoria	Rotating	33	16(P)

Canada (2013)			
Korean Advanced Institute of Science and Technology Korea (2013)	Reciprocating	20	14.1(P)
Chinese Academy of Science China (2013)	Rotating	7.9/14.9	15(P)
University of Salerno Italy (2014)	Rotating	13.5	12.5(P)
Technical University of Denmark Denmark (2015)	Rotating	10.2	12(P)
Institute of Non Ferrous Metals Poland (2016)	Reciprocating	2.5	8(P)
Federal University of Santa Catarina Brasil (2016)	Rotating	7.1	15(P)
Benedict *et al.* (2016)	Rotating	21	15(P)

* *P=Permanent Magnet, S=Superconducting magnet, E=Electromagnet*

3. Perovskite manganites

Mixed-valence perovskite manganites structures have been synthesized and examined since the second half of the twentieth century due to their superior structural, magnetic, electrical [21,22,41,43] and magnetocaloric properties [16-36]. The magnetic and structural properties of polycrystalline mixed-valance manganites first produced by Jonker and Santen [61] in 1950 were investigated and dependence of these properties on doping concentration (*x*) was determined. The basis of magnetic properties in these oxides is defined by the double-exchange (DE) mechanism by Zener [62]. The advantages of Perovskite manganites compared to Gd and GdSiGe compounds are primarily low cost and chemical stability. Also, Curie temperature in manganites has a large temperature range, therefore manganites are a good candidate for magnetic cooling at varioues temperature, particularly around room temperature. First, the magnetocaloric properties of these oxides were examined by Morelli [40], Zhang [41] and Guo [63] and determined to be proper for technological applications, leading to further studies.

3.1   Structural and magnetic properties of manganites

In general, perovskite manganite has a chemical formula $RMnO_3$ (R rare-earth cations). With the doping of atoms with different properties and ionic radii, instead of R and Mn, perovskite manganites can exhibit a very diverse electrical and magnetic properties. In MCE studies, usually a Lanthanide group element is used as an R cation. Then, another alkaline-earth element with a valency of +2 is doped to this lanthanum-based structure. The general formula of this doped perovskite manganite structure is $R_{1-x}A_xMnO_3$. Here, A represents divalent metals (Ca, Ba, Sr, Li, Na, K, Y…) ions and R represents trivalent (La, Pr, Nd, Gd, Dy, Er…) rare-earth lanthanides elements. As shown in Fig.13, a manganite perovskite structure is in the form of a cubic lattice, where six oxygen is placed around the manganite atoms in octahedral order, and a lanthanide group of rare-earth elements is located at the center of the cubic lattice. When A element is not doped to the structure, in the case of $x = 0$, the perovskite manganite structure is $R^{3+}Mn^{3+}O_3^{2-}$ and the crystal structure is as shown in Fig.13.

For $x = 0$ ($RMnO_3$) all of the manganese in the structure are in the form of $(3d^4)$ Mn^{3+}. For $x = 1$ ($AMnO_3$) all of the manganese are in the form of $(3d^3)$ Mn^{4+}. Since electrons are localized on manganese atoms for both cases (for $x = 0$ and $x = 1$), the structures exhibit antiferromagnetic insulator properties.

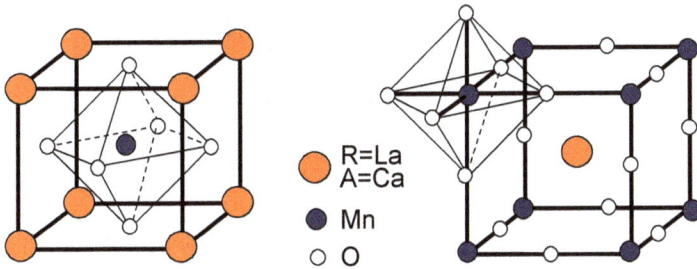

Fig.13. Schematic representation of $LaCaMnO_3$ perovskite-manganite structure.

In perovskite manganites, the difference in ionic radius of doped ions leads to structural distortions usually determined by a parameter named the Goldschmidt tolerance factor as given in Eq. (22).

$$t = \frac{(r_A + r_O)}{\sqrt{2}(r_{Mn} + r_O)} \tag{22}$$

Here, r represents ionic radious. While $t = 1$ corresponds to the ideal cubic structure with no distortions, $t < 1$ corresponds to distorted structure. The ionic radii of some ions used in the perovskite structure are given in Table 2. To understand the changes in the magnetic and electrical properties of manganites by doping atoms with different ionic radius, the magnitude of the tolerance factor is a significant parameter.

Table 2. The ionic radius of some oxides in the perovskite structure.

Ion	Radius (Å)	Ion	Radius (Å)	Ion	Radius (Å)
Al^{3+}	0.535	K^+	1.64	Rb^+	1.72
B^{3+}	0.23	La^{3+}	1.032	Sm^{3+}	1.24
Ba^{2+}	1.61	Mn^{2+}	0.83	Sn^{2+}	1.30
Bi^{3+}	1.03	Mn^{3+}	0.645	Sr^{2+}	1.44
Ca^{2+}	1.00	Mn^{4+}	0.53	Ti^{4+}	0.605
Cd^{2+}	1.31	Na^+	1.39	V^{3+}	0.74
Co^{3+}	0.61	Nd^{3+}	1.27	V^{4+}	0.63
Fe^{3+}	0.645	Ni^{3+}	0.69	Y^{3+}	0.90
Ga^{3+}	0.62	Pb^{2+}	1.49		
Gd^{3+}	1.107	Pr^{3+}	1.29	O^{2-}	1.40

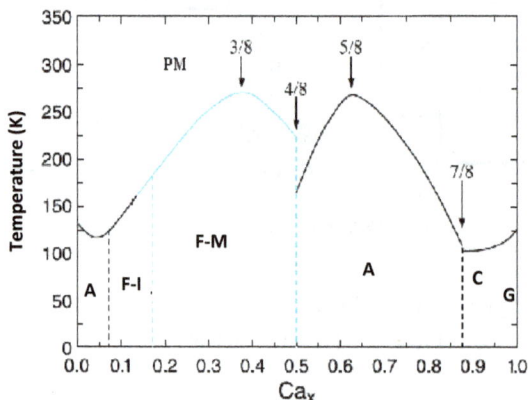

Fig.14. Magnetic phase diagram of $La_{1-x}Ca_xMnO_3$ according to the amount of (x) concentration [64].

Fig.14 displays the phase diagram of the magnetic phases in the structure for $La_{1-x}Ca_xMnO_3$ compound, depending on Ca (x) concentration. [64]. At high temperatures, at all x concentration values, the composition is in the paramagnetic insulator (PM-I) phase. At lower temperatures, depending on x, various ferromagnetic-metallic (FM-M), ferromagnetic-insulator (FM-I) and antiferromagnetic-insulator (AFM-I) phases occur. As MCE is generally observed around Curie temperature, the Curie temperature of the material is desired to be around the room temperature. As shown in Fig. 14, the maximum temperature value at which PM-FM transition is observed corresponds to the concentration ratio (260 K) $x = 0.33$.

Magnetic and electrical properties of perovskite manganite structures, depending on the changes in concentration ratio, can be explained by these changes in phase diagrams. Fig.15 shows the phase diagram for the $R_{1-x}A_xMnO_3$ structure, reflecting magnetic and electrical properties as a function of bandwidth and concentration ratio [65]. As shown in Fig.15, depending on the concentration ratio and the bandwidth, A-type AFM-I, F-type FM-M, A-type AFM-M, C-type AFM-I and G-type AFM-I like many different magnetic properties and conductivity phase is seen.

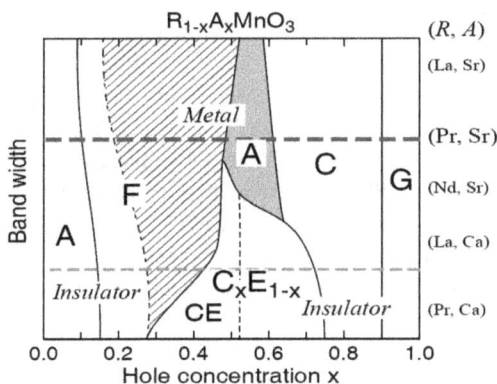

Fig.15. Phase diagram reflecting magnetic and electrical properties as a function of bandwidth and concentration ratio for $R_{1-x}A_xMnO_3$ structure [65].

3.2 Magnetocaloric properties of perovskite manganites

3.2.1 A-site substitution in manganites

3.2.1.1 (La-A)MnO₃ (A = Ca, Sr, Ba, Cd, Pb, Na, K, Ag, Bi)

The magnetocaloric properties of magnanites in $(La_{1-x}Ca_x)MnO_3$ structure have been studied extensively. It can even be called the most extensively investigated manganite group. The magnetocaloric properties of $(La_{1-x}A_x)MnO_3$ (A = Ca, Sr, Ba) manganite films were first reported by Morelli [40]. Although the obtained $|\Delta S_M|$ value is not very low, (under 1 T magnetic field change 0.5 J/kg.K) it has been shown that the peak temperature of $|\Delta S_M|$ can be adjusted in the range of 250-350 K depending on the doping cations concentration. Guo [63], revealed that the $(La_{1-x}Ca_x)MnO_3$ polycrystalline sample has a much higher $|\Delta S_M|$ value (at a field change of 1.5 T magnetic field, 5.5 J/kg.K at 230 K, 4.7 J/kg.K at 224 K and 4.3 J/kg.K at 260 K for x = 0.2, 0.25 and 0.33 samples respectively) in the concentration range of $0.20 \leq x \leq 0.33$. These values are higher than the $|\Delta S_M|$ value (4.2 J/kg.K) observed under the same magnetic field change for pure Gd [43]. Zhang [41], has shown that $La_{0.67}Ca_{0.33}MnO_3$ (x = 0.33) compound has a smaller $|\Delta S_M|$ value (0.6 J/kg.K) for 1 T magnetic field change. For the same sample, Ulyanov [68] found higher $|\Delta S_M|$ values (5.04 J/kg.K and 6.25 J/kg.K) for 0.5 T and 1 T magnetic field changes. This discrepancy among the equivalent samples can be due to different preparation methods or different chemical compositions of the samples. It has been observed that $La_{0.67}Ca_{0.33}MnO_3$ manganite has the highest $|\Delta S_M|$ value among the various x concentrations examined [15, 21, 69]. Sun [15] and Mira [21], in their study, proposed that the greater magnetic entropy change was caused by sudden decreases in magnetitation occurring in the first-order magnetic phase transition. It should not be forgotten that, with the change of the nature of the magnetic phase transition from first-order to second-order, the |SM| curve becomes more broaden even though the value of it decreases dramatically. This condition is one of the features desired for magnetic cooling [14, 21]. Hueso [70] explained that in nano-sized $La_{0.67}Ca_{0.33}MnO_{3-\delta}$ oxide, which is synthesized by the sol-gel method, peak temperature could be adjusted without repressing the magnetic entropy change. In the same study, it was noted that magnetic entropy change is proportional to grain size. Guo [71] examined the effect of grain size on magnetic entropy change in $La_{0.75}Ca_{0.25}MnO_3$ compound produced by the sol-gel method. For samples with 120 nm and 300 nm grain size, Curie temperatures were measured 177 and 224 K. It has also been observed that the magnetic entropy change has decreased with the shrinking of the grain size, but the $|\Delta S_M|$ change has become broader. Biswas [72] observed a reverse magnetocaloric effect in the polycrystalline sample of $La_{0.125}Ca_{0.875}MnO_3$ compound. This situation is attributed to the presence of a non-homogenous magnetic structure formed by the antiferromagnetic phases of the composite. Under 7 T magnetic field change, negative entropy change has been reported as -6.4 J/kg.K. In recent years, a new method, called mechanical alloying or high-energy ball milling method, has been used to produce perovskite manganites [31, 73, 74]. Studies have revealed that the ball milling method has a significant number of advantages, such as low cost, high efficiency, low-temperature synthesis and the ability to adjust grains in the desired size from micrometer to nanometer. In many studies, magnetic and magnetocaloric properties of manganites produced using high-energy ball milling method were examined.

In 2014, a study conducted by Gencer and his colleagues, the $La_{067}Ca_{033}MnO_3$ compound was produced using a high-energy ball milling method [31]. It was observed that the perovskite structure was formed for milling time of more than 4 h. When the milling time was exceeded 40 h, it was shown that the perovskite structure was completely disrupted and the amorphous structure emerged. In the 24 h milled sample, it has been reported that particle size varies from nm to a few μm. Magnetic entropy change for 12 h milled sample was measured as 0.3 J/kgK in a 6 T magnetic field. Although this value is relatively small compared to the entropy values of samples produced by other methods, magnetic entropy change has been observed to have a considerably wide temperature range. Although the $|\Delta S_M|$ value in manganites produced by ball milling method is quite low, the fact that the magnetic entropy change has an extensive temperature range makes these samples suitable magnetic refrigerant under room temperature. Bourouina [73] examined the structural, magnetic and magnetocaloric properties of the nano-particle $Pr_{0.5}Sr_{0.5}MnO_3$ compound produced using high-energy ball milling method. The samples obtained during various milling times were subjected to thermal treatment of 1400 °C for 20 h. Structural analyses have shown that all samples have tetragonal symmetry. When the milling time was over 16 h, it was observed that the average particle size was reduced to nano-size. For 4, 12 and 16 h. milled samples, Curie temperature was reported as 250, 240 and 235 K, respectively. For 5 T magnetic field change, $|\Delta S_M|$ = 2.27, 2.57, 2.58 J/kg.K and RCP = 216.33, 214.92, 204.31 J/kg were reported for 4, 12 and 16 h milled samples respectively. Again in a different study, using solid-state, sol-gel and ball milling method, $La_{0.78}Dy_{0.02}Ca_{0.2}MnO_3$ compound is produced [74]. Structural analyses have shown that all the samples have an orthorhombic structure. Also, in samples produced by a solid-state and sol-gel method, the grains were in micrometers size, smaller particle sizes were obtained for the samples produced by a ball milling method. Interestingly, the samples produced by a sol-gel method exhibited first order phase transition, while the samples produced by the other two methods showed second order phase transition. The maximum magnetic entropy changes of 1.78, 1.83 and 4.24 J/kg.K under 2 T magnetic field change were reported for samples produced by solid-state, ball milling and sol-gel methods, respectively. In the sample produced by the sol-gel method, the magnetic entropy change is much more significant than the others, which is attributed to the second-order phase transition. In spite of this, the greatest RCP value (RCP values of samples produced by solid-state, ball milling and sol-gel methods under 2 T magnetic field change,respectively; 106, 112 ve 76 J/K) was reported in the samples produced by ball milling method. The results showed that the samples produced by the high-energy ball milling method could be be promising candidates in the field of magnetic cooling.

Phan [75] studied magnetic and magnetocaloric properties in $(La_{1-x})_{0.8}Ca_{0.2}MnO_3$ (x = 0.05, 0.1, 0.2 and 0.3) compound which contains vacancies on the La side. Interestingly, La-side vacancies have not only improved the magnetocaloric properties, but also Curie temperature. Likewise, Hou [76] examined the effect of La-side vacancies on $La_{0.67-x}Ca_{0.33}MnO_3$ (x = 0, 0.02, 0.06, and 0.1) compound. The maximum $|\Delta S_M|$ value was measured as 2.78 J/kg.K for x = 0.02 sample in 277 K and 1 T magnetic field. Chen [77] examined the effect of La-side vacancies on Tc and magnetocaloric properties in $(La_{0.8-y}Ca_{0.2})MnO_3$ compound. The Curie temperature, for y = 0 value of 182 K increased up to 260 K for y = 0.05. Besides, it was observed that Tc remained almost constant for higher y values. It is noted that the nature of the magnetic phase transition is transformed into second-order for y = 0 and y = 0.01, and the first-order for y = 0.03 and 0.1. It is observed that the $|\Delta S_M|$ value in 1 T magnetic field increases from 7.7 mJ/cm^{3}K for y = 0 to 22.3 mJ/cm^{3}K for y = 0.03. In 2007, Adiguzel and his colleagues examined the effect of sintering temperature on structural, magnetic and magnetocaloric properties of $La_{067}Ca_{033}MnO_3$ compound produced using the polymeric precursor route method [26]. Magnetization measurements showed that Curie temperature increased from 241.3 K for the 600 °C sintered sample to 268.8 K for 1000 °C sintered sample. Similarly, magnetic entropy change has been found to increase due to sintering temperature. At 1150 °C sintered sample, the highest $|\Delta S_M|$ value is measured as 4.8 J/kg.K in 1 T magnetic field.

A large number of researches have studied $(La_{1-x}Sr_x)MnO_3$ manganites, to make the magnetocaloric effect available at room temperature. First, Szewczyk [78] examined the magnetocaloric properties of $La_{1-x}Sr_xMnO_3$ (x = 0.120, 0.135, 0.155, 0.185 and 0.200) manganites. In the research, the $|\Delta S_M|$ value was also increased depending on the increased Sr ratio. The highest adiabatic temperture change (ΔT_{ad}) for x = 0.2 sample was reported to be 4.15 K at 7 T magnetic field. Mira and colleagues [21], found that the $|\Delta S_M|$ value was 1.5 J/kg.K at 370 K for $La_{0.67}Sr_{0.33}MnO_3$ polycrystalline compound under 1 T magnetic field change. This value is similar to the research results obtained by Xu and his colleagues [17]. For 1 T magnetic field change, the maximum $|\Delta S_M|$ value for the $La_{0.65}Sr_{0.35}MnO_3$ sample has been reported by Phan [79] as 2.12 J/kg.K at 305 K. For the $La_{0.8}Sr_{0.2}MnO_3$ compound produced by the carbonate precursor method, under 2 T magnetic field change, $|\Delta S_M|$ value has been reported by Pekala [80] as 1.7 Jkg.K at 275 K. Immediately after this, magnetocaloric properties of the same $La_{0.8}Sr_{0.2}MnO_3$ compound, produced by the sol-gel method, were examined for the polycrystalline and nanosize forms [81]. Under 2 T magnetic field change, $|\Delta S_M|$ value has been reported, respectively, as 2.2 J/kg.K at 301 K and 0.5 J/kg.K at 295 K. Given the temperatures where $|\Delta S_M|$ maximums are observed, it is seen that these samples are promising for magnetic cooling at room temperature

First studies on magnetocaloric properties of $La_{1-x}Ba_xMnO_3$ manganites have been reported by Phan [82], Zhong [83] and Xu [17]. The magnetocaloric properties of $La_{0.7}Ba_{0.3}MnO_3$ polycrystalline manganites were first published by Phan [82]. For 1 T magnetic field change and at 336 K, $|\Delta S_M|$ value has been reported as 1.6 J/kg.K. Zhong [83] examined the effect of oxygen ratio in $La_{2/3}Ba_{1/3}MnO_{3-\delta}$ (δ = 0, 0.02, 0.05, 0.08 and 0.1) manganites on magnetic and magnetocaloric properties. With the increase of δ value, it has been observed that the value of $|\Delta S_M|$ has decreased considerably. For δ = 0, the $|\Delta S_M|$ value has been reported as 2.7 J/kg.K at 350 K in 1 T magnetic field change [83]. This value is comparatively different from the $|\Delta S_M|$ value obtained by Xu [17] for $La_{0.67}Ba_{0.33}MnO_3$ compound. It is stated that this discrepancy can be caused by differences in sample preparation and chemical compositions. Magnetic and magnetocaloric properties of $La_{1-x}Ba_xMnO_3$ (x = 0.1, 0.2 and 0.3) manganites were examined by Tonozlis [84]. It was reported that Curie temperature increased from 181 K to 319 K, depending on Ba concentration. For 2 T magnetic field change, it has been reported that the $|\Delta S_M|$ value of 1.51 J/kg.K for x = 0.1 has increased to 2.61 J/kg.K for x = 0.3. In recent years, Hussain [85] examined the magnetic and magnetocaloric properties of $La_{1-x}Ba_xMnO_3$ (x = 0.30, 0.35 and 0.40) manganites. For $La_{0.7}Ba_{0.3}MnO_3$, under 2.5 T magnetic field, they reported the $|\Delta S_M|$ value as 2.06 J/kg.K at 342 K. The results revealed that these samples could be used as magnetic coolers around room temperature. Das [86] examined magnetocaloric properties of $La_{0.7}Ba_{0.3-z}Na_zMnO_3$ ($0 < z < 0.15$) compound. The Curie temperatures were reported as 317, 313 and 312 K for z = 0.05, 0.1 and 0.15 respectively. The largest magnetic entropy change, under 0.8 T magnetic field change was observed as 1.18J/kg.K for z = 0.05. Regaieg [87] examined the magnetic and magnetocaloric properties of $La_{0.8}Na_{0.2-x}K_xMnO_3$ ($0 \leq x \leq 0.2$) compound. In K doped samples, it was reported that the Curie temperature remained constant at around 330 K. Under 1 T magnetic field change, for x = 0 at 325 K, the $|\Delta S_M|$ value was measured as 1.32 J/kg.K. The magnetic entropy value for x = 0.2 is down to 0.91 J/kg.K at 335 K. Luong [88] examined the magnetocaloric properties of $La_{1-x}Cd_xMnO_3$ (x = 0.1, 0.2, and 0.3) manganites. Under H = 1.35 T magnetic field, $|\Delta S_M|$ value was found to be 2.88 J/kg.K at 140 K for x = 0.3 and 1.01 J/kg.K for x = 0.2 at 150 K. Also it has been reported that, there is a large difference between Tc and peak temperature of $|\Delta S_M|$. This difference was related to the non-uniform distribution of the grains in the sample. Hamad [89] examined the magnetocaloric properties of the $La_{1-x}Cd_xMnO_3$ compound. The distribution of magnetic entropy change of the $La_{1-x}Cd_xMnO_3$ compound is much more uniform than gadolinium. This is the desired characteristic for an Ericsson cycled magnetic refrigerator. Also, in $La_{1-x}Cd_xMnO_3$ compound, it was seen that temperature ranges from 150 K to room temperature could be obtained with different Cd concentrations. For this reason, $La_{1-x}Cd_xMnO_3$ compounds make it possible to be used as a magnetic cooler in various temperature ranges.

A significant number of studies have been carried out on the understanding of magnetocaloric properties of $La_{1-x}Pb_xMnO_3$ manganites. Chau [90] examined magnetocaloric properties of $La_{1-x}Pb_xMnO_3$ (x = 0.1, 0.2, 0.3, 0.4 and 0.5) compounds and reported that the $|\Delta S_M|$ value increased with increasing Pb ratio up to x = 0.3 and then decreased at higher Pb values. The maximum $|\Delta S_M|$ value for the x = 0.3 sample at 1.35 T and 358 K was 1.53 J/kg.K. In another study [91] the measured $|\Delta S_M|$ and ΔT_{ad} values in $La_{1-x}Pb_xMnO_3$ (x = 0.1, 0.2, 0.3) compounds and observed the maximum $|\Delta S_M|$ value for x = 0.2 sample. Under 1.5 T magnetic field, ΔT_{ad} values were reported as 0.68 K at 292 K for x = 0.2, and 1 K at 349 K for x = 0.3. Tozri [92], reported the $|\Delta S_M|$ value, for $La_{0.8}Pb_{0.1}MnO_3$ sample, under 1 T magnetic field as; 0.43 J/kg.K at 201 K.

For $La_{1-x}Na_xMnO_3$ compound, Zhong [20, 93] showed that the $|\Delta S_M|$ peak temperature could be adjusted within the range of 195-334 K. On the other hand, for $La_{1-x}K_xMnO_3$ compound, it was shown that the peak can be adjusted in the range of 230-334 K. For $La_{1-x}Na_xMnO_3$ compound, the $|\Delta S_M|$ values were reported as 1.32, 1.53, 2.11 and 1.96 J/kg.K for x = 0.075, 0.1, 0.165 and 0.2 samples respectively under 1 T magnetic field change [20]. Das [94] later examined the magnetocaloric properties of $La_{1-x}K_xMnO_3$ (x = 0.05, 0.1, 0.15) compound produced by the pyrophoric method. In this study, it was explained that adding potassium increases Curie temperature in the structure to 264 K for x = 0.05 and 310 K for x = 0.15. At the same time the highest $|\Delta SM|$ value, under 1 T magnetic field change for $La_{0.85}K_{0.15}MnO_3$ has been reported as 3 J/kg.K at 310 K. Again, the same author [95] examined the magnetic and magnetocaloric properties of the $La_{1-x}K_xMnO_3$ compound produced in nano-size by using the pyrophoric method. It has been found that the Curie temperature varies between 260 and 309 K, depending strictly on the K ratio. Magnetic entropy changes and adiabatic temperature changes have been detected to increase proportionally with the K ratio. It has been reported that maximum $|\Delta S_M|$ and ΔT_{ad} values for $La_{0.85}K_{0.15}MnO_3$ compound, under 1 T magnetic field change, are 3 J/kg.K and 2.1 K respectively. In 2011, Juan [96] examined the relationship between magnetocaloric properties and degree of calcination in the nanoparticle $La_{1-x}K_xMnO_3$ compound. In $La_{0.85}K_{0.15}MnO_3$ compound, for the samples calcined at 600 °C, 800 °C and 1000 °C, $|\Delta S_M|$ values have been reported as 2.02, 3.06 and 3.56 J/kg.K respectively at 274 K under 2 T magnetic field change. Tang [97] found that $La_{1-x}Ag_xMnO_3$ ($0 \leq x \leq 0.3$) compounds have a huge magnetocaloric effect. In $La_{0.8}Ag_{0.2}MnO_3$ compound, for 1 T magnetic field exchange, the $|\Delta S_M|$ value has been reported as; 3.4 J/kg.K. This value is higher than the value of the Gd element. Coşkun [98] examined the magnetic and magnetocaloric properties of $La_{1-x}Ag_xMnO_3$ ($0.05 \leq x \leq$

0.25) compound. The Curie temperature increased from 200 K for $x = 0.05$ to 290 K for $x = 0.25$, depending on the concentration of Ag. Under 3 T magnetic field change, relative cooling power for $x = 0.1$, 0.15 and 0.25, were reported as; 82.49, 82.61 and 127.37 J/kg respectively. Gamzatov [99] studied the magnetocaloric properties of $La_{0.9}Ag_{0.1}MnO_3$, $La_{0.8}Ag_{0.2}MnO_3$, and $La_{0.8}Ag_{0.15}MnO_3$ compounds. When the $La_{0.8}Ag_{0.15}MnO_3$ compound is sintered at 1373 K, the $|\Delta S_M|$ value has been reported as 2.8 J/kg.K at 270 K under 2.6 T magnetic field change. Aliev [100] examined the magnetocaloric properties of $La_{1-x}K_xMnO_3$ ($x = 0.05$, 0.1, 0.11, 0.13, 0.15, 0.175) compounds. The highest adiabatic change was reported as 2.05 K and 1.66 K for $La_{0.87}K_{0.13}MnO_3$ and $La_{0.85}K_{0.15}MnO_3$ compounds under 1 T magnetic field change.

In 2014, İzgi and his colleagues [30] examined the magnetic and magnetocaloric properties of $La_{0.94}Bi_{0.06}MnO_3$ compound in detail. The structural distortion observed in $La_{0.94}Bi_{0.06}MnO_3$ compound from the ideal cubic lattice to orthorhombic lattice was related to the effect of Jahn-Teller and the polarized $6s^2$ lone-pair characteristic of Bi^{3+} ions. Magnetization measurements revealed that small amounts ($x = 0.06$) of Bi doping in $LaMnO_3$ compound caused ferromagnetic order. For the sample, considerably high magnetic entropy change ($|\Delta Sm| = 1.58$ J/kg.K) has been reported at 209 K under 1 T magnetic field change. In 2015, Kolat and colleagues [34] systematically examined the magnetocaloric properties of $La_{1-x}Bi_xMnO_3$ ($x = 0.01$, 0.03, 0.06, 0.1 and 0.2) compounds. The Curie temperatures reported as 234, 224, 209, 198, 149 K for $x = 0.01$, 0.03, 0.06, 0.1, 0.2 samples respectively. Saturation magnetization has been reported as; 89, 88, 85, 84, 80 emu/g, respectively. Ferromagnetism observed in $La_{1-x}Bi_xMnO_3$ compounds is explained by ferromagnetic super-exchange interaction between Mn^{3+}–O–Mn^{3+} ions. It has been reported that the $|\Delta Sm|$ value, under 1 T magnetic field change, has decreased from 2.42 J/kg.K for $x = 0.01$ to 0.79 J/kg.K for $x = 0.2$. These $|\Delta Sm|$ values obtained are comparable to the magnetic entropy change observed for many manganites. For example, 1.58 J/kg.K value observed for $x = 0.06$, is comparable to the value [101] of 1.55 J/kg.K measured for $La_{0.67}Sr_{0.33}MnO_3$ compound in the same magnetic field, and the value of 1.6 J/kg measured in the $La_{0.67}Ba_{0.33}MnO_3$ compound [102].

3.2.1.2 La(Ca-A')MnO₃ (A' = Sr, Ba, Pb, K, Na, Ag, Mg)

As discussed in detail in section 3.2.1.1, $La_{1-x}Ca_xMnO_3$ compounds exhibited the highest magnetocaloric effect among existing manganites [15, 21, 26, 31, 40-43, 56, 59, 60-64]. However, since Curie temperatures are below room temperature, where the magnetic entropy change occurs in general, it limits the use of $La_{1-x}Ca_xMnO_3$ compounds as a magnetic refrigerant at room temperature. For this reason, new dopings are being investigated that raise Curie temperature around room temperature without altering the high magnetic entropy change in $La_{1-x}Ca_xMnO_3$ compounds. In this context, a large number of studies have been carried out on the replacement of Ca element by Sr, Ba, Pb, K, Na, Ag. Phan [14, 103] reported a relatively high magnetocaloric effect in a monocrystalline $La_{0.7}Ca_{0.3-x}Sr_xMnO_3$ (x = 0.05, 0.10, 0.20 and 0.25) compound, around room temperature. For x = 0.05, under 5 T magnetic field change and at 275 K, the $|\Delta S_M|$ value measured as 10.5 J/kg.K. This value is more significant than the magnetic entropy change of the Gd element. Therefore, single crystalline manganite is one of the important candidate for magnetic cooling at room temperature. Again, a lot of studies have been carried out on the polycrystalline samples of the same example. In these studies, it was observed that there was a decrease in magnetic entropy change as a result of the increase in Sr ratio. Sun [104] reported the $|\Delta S_M|$ value at 315 K for 2 T magnetic field change in the compound of $La_{0.7}Ca_{0.2}Sr_{0.1}MnO_3$ as 2.85 J/kg.K. Again, Li [105] reported the $|\Delta S_M|$ value, at 317 K for 2 T magnetic field change in the $La_{0.5}Ca_{0.3}Sr_{0.2}MnO_3$ compound, as 1.52 J/kg.K. In a study conducted by Gou [22], it was reported that there was a structural phase transition in $La_{0.75}Ca_{0.25-x}Sr_xMnO_3$ compound depending on the Sr ratio. For $x \leq$ 0.125, the composition has an orthorhombic phase, while for $x \geq$ 0.125 it has a rhombohedral phase [22]. Kim [106] examined the magnetic and magnetocaloric properties of $La_{0.7}Ca_{0.3-x}Sr_xMnO_3$ (x = 0.120, 0.135 and 0.150) compound to elucidate the effect of lattice structure on T_c and magnetocaloric properties. For x = 0.135, the magnetization curve observed two-phases reduction in the values 309 and 320 K corresponding to the Curie temperatures of orthorhombic and rhombohedral phases, while the samples x = 0.12 and 0.15 exhibit the typical magnetization behavior at Curie temperatures of 300 and 323 K. Also for x = 0.12, 0.13 and 0.15, the $|\Delta S_M|$ value measured as; 1.87, 1.72 and 1.7 J/kg.K , respectively. Studies have revealed that the increase in Sr ratio has a positive effect on Curie temperature, especially when shifting from orthorhombic to rhombohedral phase, although it has a sharp increase in Curie temperature, it has an adverse impact on $|\Delta S_M|$ [14,22,101-106]. Mira [21] conducted a comprehensive study of the magnetic and magnetocaloric properties of $La_{2/3}(Ca_{1-x}Sr_x)_{1/3}MnO_3$ (x = 0, 0.05, 0.15, 0.25, 0.50, 0.75 and 1) compounds. Depending on the Sr ratio, the decrease in $|\Delta S_M|$ value is attributed to the change like the magnetic phase transition. In the study, the sample x = 0.15 showed the first-order magnetic phase transition. For Sr, it was observed that the nature of phase transitions changed from first to second order with higher content of Sr. Phan [82] examined magnetocaloric properties of $La_{0.7}Ca_{0.3-x}Ba_xMnO_3$ (x = 0.12, 0.24 and 0.3) compounds and reported that the $|\Delta S_M|$ value decreased by increasing Ba content. The $|\Delta S_M|$ values were reported as 1.85 J/kg.K at 298 K for x = 0.12, 1.72 J/kg.K at 320 K for x = 0.24 and 1.6 J/kg.K at 336 K for x = 0.3 under 1 T magnetic field change. Sun [107] examined the magnetocaloric properties of $La_{2/3}(Ca,Pb)_{1/3}MnO_3$ compound and $|\Delta S_M|$ value for 7 T magnetic field change is reported as; 7.5 J/kg.K at 290 K and ΔT_{ad} value is reported as 5.6 K. The value of $|\Delta S_M|$ for $La_{2/3}(Ca,Pb)_{1/3}MnO_3$ was lower than the value of $|\Delta S_M|$ for $La_{2/3}Ca_{1/3}MnO_3$ compound under the same field change [102]. Phan [108] studied the magnetocaloric properties of $La_{0.6}Ca_{0.3}Pb_{0.1}MnO_3$, $La_{0.7}Ca_{0.2}Pb_{0.1}MnO_3$, and $La_{0.7}Ca_{0.1}Pb_{0.2}MnO_3$ compounds. The maximum $|\Delta S_M|$ value under 1.35 T magnetic field change for $La_{0.7}Ca_{0.1}Pb_{0.2}MnO_3$ has been reported as; 3.72 J/kg.K. at 337 K. Hanh [109] examined magnetocaloric properties of $La_{0.7}Ca_{0.3-x}Pb_xMnO_3$ (x = 0.05, 0.01, 0.15 and 0.2) compounds. Under the magnetic field change of 1.35 T, for x = 0.05 and 0.2 samples, respectively; at 270 and 337 K, the same $|\Delta S_M|$ value (3.72 J/kg.K) reported. The magnetocaloric properties of $La_{0.7}Ca_{0.3-x}K_xMnO_3$ (x = 0.05, 0.075 and 0.1) polycrystalline perovskites were examined by Bejar [110]. $|\Delta S_M|$ values has been reported as 3.95 J/kg.K at 270 K for x = 0.05, 3.75 J/kg.K at 281 Kfor x = 0.075 and 3.49 J/kg.K at 272 K for x = 0.1, respectively under 2 T magnetic field change. Koubaa [111] examined the magnetic and magnetocaloric properties of $La_{0.65}Ca_{0.35-x}Na_xMnO_3$ compound, and it has been reported that the Curie temperature increased from 248 K for x = 0 to 315 K for x = 0.2, with the increase of Na ratio. Under 5 T magnetic field change, the maximum $|\Delta S_M|$ value has been reported as; 3 J/kg.K for x = 0.05 and 5.8 J/kg.K for x = 0.2. The magnetic and magnetocaloric properties of $La_{0.5}Ca_{0.5-x}Na_xMnO_3$ compound were studied by Mehri [112], and he reported that with the increasing Na ratio, Curie temperature decreased and magnetic entropy change increased. The same author [113] examined the magnetic and magnetocaloric properties of the $La_{0.5}Ca_{0.5-x}Ag_xMnO_3$ compound and observed characteristics similar to those found in Na-doped perovskites. In recent years, Kolat [33] studied the magnetic and magnetocaloric properties of $La_{0.67}Ca_{0.33-x}Mg_xMnO_3$ (x = 0, 0.02, 0.05, 0.1, 0.2, 0.33) compounds. It has been reported that the Curie temperature decreases from 267 K for x = 0, to 96 K for x = 0.33, along with the amount of Mg. The reason for the decrease in Curie temperature and similarly in saturation magnetism was attributed to the weakening of ferromagnetism in the Mg-doped samples. Under 1 T magnetic field change, $|\Delta S_M|$ value is decreased from 4.07 J/kg.K for x = 0 to 0.41 J/kg.K for x = 0.33. This reduction in magnetic entropy change is attributed to the decrease in the saturation magnetization and the transformation of the nature of the magnetic phase transition from first order to second order. As can be seen from the results, combining A'= Sr, Ba, Pb, K, Na, Ag, Mg compounds instead of Ca in La(Ca–A')MnO$_3$ compound causes an increase in Curie temperature and decrease in magnetic entropy change in general. With the determination of appropriate doping elements, manganites can be produced which provides ample magnetic entropy change at suitable temperatures and can be used as a magnetic cooler at room temperature.

3.2.1.3 La(Sr-A')MnO$_3$ (A' = Ba, K, Ag, Mg)

Phan [79], reported considerably sizeable magnetic entropy change, $|\Delta S_M|$ = 2.26 J/kg.K at 354 K under 1 T magnetic field change in the La$_{0.6}$Sr$_{0.2}$Ba$_{0.2}$MnO$_3$ compound. Koubaa [114] examined magnetocaloric properties of La$_{0.7}$Sr$_{0.3-x}$Ag$_x$MnO$_3$ (x = 0.05, 0.1, 0.15 and 0.2) composites and observed that the Curie temperature decreased from 365 K to 286 K with the increase of Ag element from x = 0 to x = 0.2. The most considerable magnetic entropy change, for La$_{0.7}$Sr$_{0.2}$Ag$_{0.1}$MnO$_3$ compound was reported as 0.9 and 4.5 J/kg.K under 1 and 7 T magnetic field changes respectively. Again, the magnetocaloric properties of La$_{0.7}$ Sr$_{0.3-x}$K$_x$MnO$_3$ (x = 0.05, 0.1, 0.15 and 0.2) compounds have been examined by the same scientist [115]. In present samples, it was observed that the Curie temperature decreases from 365 K for x = 0, to 328 K for x = 0.2, depending on the K value. $|\Delta S_M|$ values were reported as 1.37 and 1.2 J/kg.K for x = 0.05 and 0.15 samples at 1 T magnetic field change. Wang [116] studied the magnetic and magnetocaloric properties of La $_{0.67}$Sr$_{0.33-x}$Mg$_x$MnO$_3$ (x = 0, 0.05, 0.15 and 0.2) compounds. With the increase of Mg content, both Curie temperature and saturation magnetization has been reported to decrease. Under 5 T magnetic field change, the $|\Delta S_M|$ value for x = 0, 0.05, 0.15, 0.2, measured as; 2.49, 1.28, 1.37 and 1.3 J/kg.K respectively. Mg-doped manganites have shown almost a constant magnetocaloric effect, in a considerably wide temperature range (50 K to 300 K), which is desirable for magnetic cooling.

3.2.1.4 (La-A)CaMnO$_3$ (A = Nd,Tb, Dy, Gd, Ce, Y, Sm, Bi, Eu, Ho)

Many studies have been carried out on the (La−A)CaMnO$_3$ group of manganites, and it has been shown that magnetocaloric properties can be improved by substitution of the La element with elements such as Nd, Bi, Tb, Dy, Gd, Ce, Y, Sm, Eu, Pr, Ho. Wang [117] examined the influence of the substitution of La element with Nd element on magnetocaloric properties in La$_{0.7-x}$Nd$_x$Ca$_{0.3}$MnO$_3$ (x = 0, 0.05, 0.1, 0.15 and 0.20) compounds. Highest $|\Delta S_M|$ value, under 1 T magnetic field change, has been reported for x = 0.2 as 2.31 J/kg.K at 213 K. Chen [16] examined the magnetocaloric properties of (La$_{1-x}$R$_x$)$_{2/3}$Ca$_{1/3}$MnO$_3$ (R = Gd, Dy, Tb, Ce, x = 0–0.2) compounds and they observed that Curie temperature decreased due to partial replacement of the La with Gd, Dy, and Tb elements. Interestingly, the maximum $|\Delta S_M|$ value for all doping elements was observed in x = 0.1. For (La$_{0.9}$Dy$_{0.1}$)$_{2/3}$Ca$_{1/3}$MnO$_3$ component, under 1.5 T magnetic field change and at 175 K, the highest $|\Delta S_M|$ value has been reported as 6.06 J/kg.K. Zhang [41] pointed that the partial replacement of La with Y element in La$_{0.60}$Y$_{0.07}$Ca$_{0.33}$MnO$_3$ compound, reduces both magnetocaloric properties and the Curie temperature. The $|\Delta S_M|$ value measured as 1.46 J/kg.K at 230 K under 3 T magnetic field change. Anwar [118] examined the effect of Sm doping at different ratios instead of La, on magnetic and magnetocaloric properties in $La_{0.7-x}Sm_xCa_{0.30}MnO_3$ ($0 \leq x \leq 0.3$) compounds. From magnetization and Arrott analysis, it was concluded that the sample exhibited a first-order ferromagnetic phase transition for $x = 0$ and Sm doped samples exhibited a second-order ferromagnetic phase transition. Curie temperature was reported to decrease from 182 K for $x = 0.05$ to 109 K for $x = 0.3$. Although the change in magnetic entropy (1.75 J/kg.K under 1 T magnetic field) is quite significant for $x = 0$, the $|\Delta S_M|$ value has decreased considerably due to irregularities caused by Sm in Sm-containing samples. Another reason for the reduction in magnetic entropy change is that the nature of the phase transition transformed from -first to second-order. Zhang [119] examined the magnetocaloric properties of $La_{0.65-x}Eu_xCa_{0.35}MnO_3$ ($x = 0$, 0.05, 0.10 and 0.15) compounds prepared by sol-gel method. With the increasing Eu content, Curie temperature has decreased. The most substantial magnetic entropy exchange, under 1.5 T magnetic field change, has been reported as 5.778 J/kg.K for $La0_.6Eu_{0.05}Ca_{0.35}MnO_3$ compound. Ning [120] examined the magnetic and magnetocaloric properties of $(La_{0.8}Ho_{0.2})_{2/3}Ca_{1/3}MnO_3$ and $(La_{0.5}Ho_{0.5})_{2/3}Ca_{1/3}MnO_3$ compounds nanoparticles ranging from 50 to 200 nm. prepared by the sol-gel method. Under 5 T magnetic field change, for $(La_{0.8}Ho_{0.2})_{2/3}Ca_{1/3}MnO_3$ at 100 K, $|\Delta S_M|$ has been reported as 1.19 J/kg.K, and for $(La_{0.5}Ho_{0.5})_{2/3}Ca_{1/3}MnO_3$ at 152 K, $|\Delta S_M|= 2.03$ J/kg.K. Anwar [121] examined the magnetic and magnetocaloric properties of $La_{(0.75-x)}Ce_xCa_{0.25}MnO_3$ ($x = 0.0$, 0.2, 0.3 and 0.5) compounds and reported that Curie temperature decreased due to Ce concentration (255, 213 and 150K for $x = 0$, 0.2 and 0.3 respectively). Under the magnetic field changes of 1.5 and 4 T, the most considerable magnetic entropy change has been reported as 3.31 and 6.4 J/kg.K for $La_{0.55}Ce_{0.2}Ca_{0.25}MnO_3$. Gencer [23] examined the effect of Bi doping on sintering temperature and magnetic and magnetocaloric properties of $La_{0.62}Bi_{0.05}Ca_{0.33}MnO_3$. Interestingly, the result of Bi dopings on magnetic and magnetocaloric properties has been determined to have a positive effect on sintering temperature as well. It has been observed that even a small Bi ratio ($x = 0.05$) decreases the sintering temperature by 200 °C. The Bi-doped sample also exhibited a remarkably high magnetic entropy value. Under 1 T magnetic field and at 248 K, this value has been reported as 3.5 J/kg.K. This value is higher than that of initial sample of $La_{0.67}Ca_{0.33}MnO_3$ compound. Although magnetic entropy change is high enough, the decrease in Curie temperature restricts the use of these samples as magnetic refrigerant at room temperature. Atalay [24] systematically studied the effect of Bi dopings on the magnetic and magnetocaloric properties of $La_{0.67-x}Bi_xCa_{0.33}MnO_3$ ($x = 0$, 0.05, 0.1, 0.2) compounds. The Curie temperatures were reported as 267, 248, 244 and 229 K for $x = 0$, 0.05, 0.1 and 0.2 respectively. The maximum magnetic entropy change was reported as 6.08 J/kg.K for the $x = 0.05$ sample at 3 T magnetic field change. The most significant decrease in magnetic entropy change was observed in the $x = 0.2$ sample. Gutierrez [122] studied the magnetocaloric properties of $(La_{0.55}Bi_{0.15})Ca_{0.3}MnO_3$, compound. Curie temperature has been reported to be 230 K. Neutron diffraction experiments have revealed that under this temperature there are localized antiferromagnetic phases within the ferromagnetic phase of the sample. Under 9 T magnetic field change, $|\Delta S_M| = 1.1$ J/kg.K and $\Delta T_{ad} = 2.3$ K were reported. In $La_{0.7}Ca_{0.3}MnO_3$ compound, as a result of the substitution of La with different elements, magnetic entropy change in general increases or remains constant, while Curie temperature is generally observed to decrease below room temperature. This situation shows that the above-mentioned manganites may be a good magnetic refrigerant candidate in the temperature range of about 210-270 K, while limiting its use as a magnetic refrigerant at room temperature.

3.2.1.5 (La-A)SrMnO₃ (A = Er, Eu, Gd, Ce, Pr, Nd, Bi)

Amaral [123] revealed that when the La ions were replaced by Er and Eu ions in $La_{0.7-x}Er_xSr_{0.3}MnO_3$ ($x = 0.014, 0.035, 0.14$ and 0.21) and $La_{0.7-x}Eu_xSr_{0.3}MnO_3$ ($x = 0.035, 0.14$ and 0.21) compounds, the Curie temperature decreases to around room temperature, whereas the magnetic entropy change remained constant. Bouderbala [124] observed that $La_{0.7-x}Eu_xSr_{0.3}MnO_3$ ($x = 0, 0.1, 0.2$ and 0.3) polycrystalline manganites exhibited structural phase transitions from rhombohedral symmetry to orthorhombic symmetry for $x \geq 0.1$. It is seen that the Curie temperature decreases from 343 K for $x = 0.1$ to 272 K for $x = 0.3$, depending on Eu concentration. All samples have been reported to exhibit high magnetic entropy change at a level that can be used as a magnetic refrigerant around room temperature. Similarly, Sudharshan [125] studied the magnetic and magnetocaloric properties of $La_{0.7-x}Eu_xSr_{0.3}MnO_3$ ($x = 0.0, 0.1, 0.2, 0.3$) compounds. Similar to previous studies, the crystalline structure has been transformed from the rhombohedral phase for $x = 0$ to orthorhombic phase for $x \geq 0.1$. Under 6 T magnetic field change, the magnetic entropy change was reported to increase from 3.88 J/kg.K for $x = 0$ to 5.03 J/kg.K for $x = 0.03$. The magnetocaloric properties of $La_{0.7-x}Pr_xSr_{0.3}MnO_3$ ($x = 0, 0.1, 0.2, 0.3, 0.4, 0.5, 0.6, 0.7$) compounds were measured by Gamzatov [126], using direct and indirect methods. Magnetic entropy change was reported to vary in the range of 1.84 and 4.21 J/kg.K under 1.8 T magnetic field change. The adiabatic temperature change (ΔT_{ad}) was measured between 1.09 and 1.75 K. Phromchuai [127] examined the magnetocaloric properties of the $La_{0.75-x}Gd_xSr_{0.25}MnO_3$ ($x = 0-0.3$) compounds prepared using the sol-gel method. Although Curie temperature decreases with increasing Gd content, the magnetic entropy change was reported to increase from 0.93 J/kg.K for $x = 0$ to 1.14 J/kg.K for $x = 0.3$ under 0.7 T magnetic field change. A significant magnetic entropy change was first reported by Kallel [128] in the Ce-doped $La_{0.7}Sr_{0.3}MnO_3$ compound. For $(La_{0.56}Ce_{0.14})Sr_{0.3}MnO_3$ compound, the magnetic entropy change has been reported as 1.55 and 4.78 J/kg.K at 357 K under 1 and 5 T magnetic field changes respectively.

Anwar [129] reported that the Curie temperature decreased from the 370 K for $x = 0$ to 310 K for $x = 0.3$ in $La_{0.7-x}Ce_xSr_{0.3}MnO_3$ ($0 \leq x \leq 0.3$) compounds, depending on the Ce concentration. It has been observed that the magnetic entropy change increases with the increasing Ce ratio up to $x = 0.15$. Under 2 T magnetic field change, for $La_{0.55}Ce_{0.15}Sr_{0.3}MnO_3$ compound, the maximum magnetic entropy change has been reported as 2.12 J/kg.K at 356 K. Çetin [130] noted that the Curie temperature decreased from 358 K for $x = 0$ to 286 K for $x = 0.3$ in $(La_{1-x}Sm_x)_{0.67}Pb_{0.33}MnO_3$ ($x = 0, 0.1, 0.2, 0.3$) compounds, depending on Sm concentration. Under 3 T magnetic field change, the adiabatic temperature change (ΔT_{ad}) is measured as 1.3 K for $x = 0.3$. Dhahri [131] investigated the magnetic entropy chage in $La_{0.7-x}Eu_xBa_{0.3}MnO_3$ ($x = 0.05, 0.1$ and 0.15) compounds and $|\Delta S_M|$ was reported to be 2.3 J/kg.K at 298 K under 1 T magnetic field change for $x = 0.15$. Barik [132] examined the effect of Bi dopings on the magnetic and magnetocaloric properties of $La_{0.7-x}Bi_xSr_{0.3}MnO_3$ ($x = 0.0$–0.4) compounds. It is observed that the Curie temperature decreases from 365 K for $x = 0$ to 191 K for $x = 0.3$. Under 5 T magnetic field change, it has been observed that the magnetic entropy change increases from 4.56 J/kg.K for $x = 0$ to 5.02 J/kg.K for $x = 0.05$. With an increasing Bi ratio, $|\Delta S_M|$ has decreased again (3.1 J/kg.K for $x = 0.3$).

3.2.1.6 $(A_{1-x}A'_x)MnO_3$ (A = Nd, Pr, Sm, Gd, Na, Eu A' = Ca, Sr, Pb, Bi)

In recent years, studies on the magnetocaloric properties in manganites with charge-order (CO) state have begun to increase. The most characteristic feature of CO manganites is to exhibiting successive two phase transitions. The first corresponds to first-order transition from the antiferromagnetic to the ferromagnetic phase at lower temperatures. The second corresponds to second-order transition from ferromagnetic metallic to paramagnetic insulator phase at higher temperatures [133]. CO state is an interaction that is based on the localization of charge carriers and naturally competes with the ferromagnetic double-exchange interactions. CO in manganites is a form of configuration in which transition metals (Mn^{3+} and Mn^{4+}) with different oxidation state form similar to the arrangement of checkers. This configuration usually causes charges to be localized and prevents electrons from hopping from one ion to another. This provides the structure of a semiconductor or insulator character. CO configuration in manganites usually occurs in the case of mixed-valence state especially for $x = 1/8, 1/2$ and $3/4$ ratios.

Si and colleagues [134] examined magnetocaloric properties in $Nd_{2/3}Sr_{1/3}MnO_3$ compound, under 1 T magnetic field change and at 257.5 K the $|\Delta S_M|$ value has been reported as 3.25 J/kg.K. In $Nd_{1-x}Sr_xMnO_3$ ($x = 0.3, 0.5$) single crystalline compounds, ΔT_{ad} and $|\Delta S_M|$ values were reported to be 20 K and 3.82 J/kg.K at 203 K under 1.4 T magnetic field change for $x = 0.3$ ($Nd_{0.7}Sr_{0.3}MnO_3$) sample [135]. For charge-ordered

$Nd_{0.5}Sr_{0.5}MnO_3$ sample, a considerable large magnetic entropy change was observed under quite small magnetic field change (1.4 T) and around the Neel temperature of 150 K [135]. Phan [136] showed that the $(Nd_{1-x}Y_x)_{0.7}Sr_{0.3}MnO_3$ ($x = 0$ and 0.07) compounds exhibited first-order phase transition for x=0 and second-order phase transition for $x = 0.07$. The Curie temperature was reported as 240 K for $x = 0$ and 170 K for $x = 0.07$. For the studied samples; under 5 T magnetic field change, magnetic entropy change was measured as 8 J/kg.K, and the relative cooling power RCP was determined in the range of 200-246 J/kg. Beiranvand [137] examined the magnetic and magnetocaloric properties of $Gd_{1-x}Ca_xMnO_3$ and $Nd_{1-x}Ca_xMnO_3$ ($0 \leq x \leq 1$) compounds. The highest magnetocaloric properties were observed in both samples at temperatures below 140 K and for low x values. Low magnetic hysteresis and high entropy change have made these materials promising candidate as a magnetic refrigerants at low temperatures. In the study for $R_{0.15}Ca_{0.85}MnO_3$ (R = Y, Gd and Dy) compound, Dhal [139] demonstrated that compounds exhibit an antiferromagnetic phase transition at Neel temperatures of 111, 119, and 112 K for R = Y, Gd, and Dy respectively. The magnetocaloric effect was calculated from isothermal magnetization curves, and the inverse magnetocaloric effect was seen at the antiferromagnetic transition temperature. Magnetocaloric properties of the $Nd_{0.5}Sr_{0.5}MnO_3$ compound were first reported by Sande [139]. Under 1 T magnetic field change and at $T_{CO} = 155$ K, a considerable magnetic entropy change, such as 2.8 J/kg.K, has been observed. Besides, it has been found that the measured $|\Delta S_M|$ value around the first-order phase transition is approximately three times higher than the observed value around the second-order phase transition. This is related to the suppression of CO-configuration as a result of increasing electron mobility under the applied magnetic field. Chen [133] reported a much larger $|\Delta S_M|$ =7.5 J/kg.K value at 183 K, under 1 T magnetic field change. In 2010, Fan [140] reported a considerable decrease in the Curie temperature and magnetic entropy value of $Nd_{0.5}Ca_{0.25}Sr_{0.25}MnO_3$, compound. The $|\Delta S_M|$ value was measured as 0.77 J/kg.K at 175 K under 1 T magnetic field change. The same author [141] reported the value of $|\Delta S_M| = 3.14$ J/kg.K, under 1.5 T magnetic field for $Nd_{0.6}La_{0.1}Sr_{0.3}MnO_3$ compound. Nanto [142] stated that $Nd_{0.5}Sr_{0.5}MnO_3$ single crystalline sample was exhibited antiferromagnetic CO phase transition at the temperature of $T_{CO} = 152$ K. On the other hand, he reported that the sample showed a ferromagnetic, paramagnetic phase transition at $T_c = 272$ K. Maximum magnetic entropy change, $|\Delta S_M|$, has been reported as 1.65 J/kg.K in first-order phase transition and -1.13 J/kg.K in the second-order phase transition. Cao [143] examined the magnetocaloric properties of $Eu_{1-x}Sr_xMnO_3$ ($x = 0.5, 0.6, 0.7$ and 0.8) compounds systematically. Under 5 T magnetic field change, the most considerable magnetic entropy change measured as 0.31 J/kg.K for $Eu_{0.5}Sr_{0.5}MnO_3$ sample. It was observed that the magnetic entropy change of the $Eu_{0.5}Sr_{0.5}MnO_3$, which was subjected to thermal treatment at 800 °C again for 12 h, increased to the value of 3.03 J/kg.K.

Ayaş [144] studied the effect of Pr doping on the magnetic and magnetocaloric properties in $(La_{1-x}Pr_x)_{0.85}Ag_{0.15}MnO_3$ ($0.0 \leq x \leq 0.5$) compounds. Rietveld analysis revealed that the compound has a rhombohedral phase till $x \leq 0.2$, but orthorhombic phase for $x \geq 0.3$. It has also been observed that the average particle size is reduced depending on the Pr concentration. All samples exhibit a second-order magnetic phase transition. The Curie temperature decreases from 262 K to 138 K, depending on the increase of Pr content. Under 5 T magnetic field change, magnetic entropy change and relative cooling power were measured in the range of 7.9-2.88 J/kg.K and 213.32-153.5 J/kg, respectively.

In 2000, Chen [145] examined magnetocaloric properties of $Pr_{1-x}Sr_xMnO_3$ ($x = 0.3$, 0.4, and 0.5) polycrystalline manganites, and they measured the maximum $|\Delta S_M|$ value as 7.1 J/kg.K at 160 K under 1 T magnetic field change for $x = 0.5$. Among these samples, only $Pr_{0.5}Sr_{0.5}MnO_3$ sample exhibits CO configuration at $T_{CO} = 161$ K. Chen [146], has showed that the replacement of Pr with Nd in $(Pr_{1-y}Nd_y)_{0.5}Sr_{0.5}MnO_3$ ($y = 0$, 0.3, 0.5, 0.7 and 1.0) causes an increase in Curie temperature from 205 K to 267 K and an increase in CO transition temperature from 161 K to 183 K. For all samples, it has been observed that the $|\Delta S_M|$ value (6.5 J/kg.K for $x = 0$, 8J/kg.K for $x = 1$) remains almost constant under 1 T magnetic field change. It has been showed that the Curie temperature decreases from 310 K for $x = 0$ to 252 K for $x = 0.4$, depending on the Bi doping in the $(Pr_{1-x}Bi_x)_{0.6}Sr_{0.4}MnO_3$ ($0 \leq x \leq 0.4$) compound [147]. For $x = 0.06$, the magnetic entropy change has been reported as 1.11 J/kg.K and 4.78 J/kg.K, under 1 and 7 T magnetic field changes. In another study, Gomes [148] examined the magnetocaloric properties of $Pr_{1-x}Ca_xMnO_3$ ($0.3 \leq x \leq 0.45$) manganites and reported positive and negative magnetic entropy changes in a relatively high value. For $Pr_{0.68}Ca_{0.32}MnO_3$ sample, the $|\Delta S_M|$ value was measured as positive 24 J/kg.K at 21.7 K and negative 27 J/kg.K at 31 K, under 5 T magnetic field change. It can be revealed from the results, these samples are particularly suitable for magnetic cooling at low temperatures. The same authors [149] investigated the contribution of CO configuration to magnetic entropy change in $Pr_{1-x}Ca_xMnO_3$ ($0.2 \leq x \leq 0.95$) compounds. The value of ΔS_M observed in CO manganites was shown to caused by overlapping of negative entropy change resulting from the spin configuration (ΔS_{spin}) and the positive entropy change resulting from charge-order configuration (ΔS_{CO}). Immediately after that, Phan [150] has reported a considerably high $|\Delta S_M|$ value (8.52 J/kg.K) in single crystalline $Pr_{0.63}Sr_{0.37}MnO_3$ compound, under 5 T magnetic field change at 300 K. A relatively large magnetic entropy change in a relatively low magnetic field change is one of the characteristics that are required for the application of active magnetic coolers at home. For commercial applications of magnetic cooling at room temperature, a single crystalline sample of $Pr_{0.63}Sr_{0.37}MnO_3$ could be a promising candidate. To completely understand the properties of Pr-based samples, the same author [151] examined the magnetocaloric properties of $Pr_{1-x}Pb_xMnO_3$ ($0.1 \leq x \leq 0.5$) compound in detail. Under 1.35 T magnetic field change, $|\Delta S_M|$ value for $x = 0.1$, 0.4 and 0.5 has been reported as; 3.91, 3.68 and 3.34 J/kg.K respectively. These values are larger than the magnetic entropy change of the Gd element. More importantly, these entropy changes are obtained under a relatively low magnetic field, which can be produced by permanent magnets. Bingham [152] examined the magnetocaloric properties of the charge ordered compound $Pr_{0.5}Sr_{0.5}MnO_3$. It has been observed that the system has exhibited paramagnetic-ferromagnetic phase transitions at $T_c = 255$ K and a phase transitions from CO ferromagnetic phase to CO antiferromagnetic phase at $T_{CO} = 165$ K. The value of $|\Delta S_M| = 7.5$ J/kg.K obtained at $T_{CO} = 165$ K is two times higher than the value of $|\Delta S_M| = 3.2$ J/kg.K observed at $T_c = 255$ K under 5 T magnetic field change. In $Pr_{0.5}M_{0.1}Sr_{0.4}MnO_3$ (M = Eu, Gd and Dy) compound, the Curie temperature for Eu, Gd and Dy doped samples were measured as 270, 258 and 248 K respectively [153]. The Arrott plots have showed that all samples exhibit a second-order magnetic phase transition. Under 1 T magnetic field change, the magnetic entropy change was reported as 1.37, 1.23 and 1.18 J/kg.K for M = Eu, Gd and Dy, respectively.

Sarkar [154] studied the magnetocaloric properties of monocrystalline $Sm_{0.52}Sr_{0.48}MnO_3$ compound. At 125 K, $|\Delta S_M|$ value was reported to be 5.9 J/kg.K, under 1 T magnetic field change. In another study, it was observed that $Sm_{1-x}Sr_xMnO_3$ ($x = 0.42$, 0.44, 0.46) compounds exhibited CO configuration, for $x = 0.44$ [155]. Magnetization measurements have revealed that all samples show a first-order phase transition. Curie temperature has been reported as 130, 143, 133 K for $x = 0.42$, 0.44, 0.46 respectively. The highest magnetic entropy change was determined to be 4.61 J/kg.K for $x = 0.44$ under 5 T magnetic field change. It was found that RCP = 151.42, 140.15, 135.91 J/kg for $x = 0.42$, 0.44, 0.46 respectively under the same field change. Zashchirinskii [156] examined the ceramic structure of $Sm_{0.55}Sr_{0.45}MnO_3$ compound and the magnetocaloric properties of three different samples of the single crystal structure subjected to thermal treatment in oxygen and air condution environment. The temperatures at which the maximum adiabatic change (ΔT_{ad}) observed were reported as 143.3 K for the ceramic structure, 244 K for the single crystal sample obtained in the oxygen environment, and 143 K for the single crystal sample obtained in the air environment. At these temperature values, ΔT_{ad} changes were measured as 0.8, 0.41 and 0.4 K, respectively.

Kolat [29] investigated the metamagnetic phase transition and magnetocaloric properties in the Charge-Ordered $Pr_{0.68}Ca_{0.32-x}Sr_xMnO_3$ ($x = 0$, 0.1, 0.18, 0.26 and 0.32) compounds. In low Sr concentrations, ($x = 0$ and 0.1), magnetization curves have shown CO transition around 185 K. For further Sr concentrations, peaks representing CO transition were observed to disappear. For x = 0 and 0.1 samples, sharp step-like metamagnetic transitions were seen in magnetization curves (M-H loops). At low Sr concentrations, two different abnormal magnetic entropy changes were observed just below and above the Curie temperature. The positive ΔS_M value (0.45 J/kg.K for x = 0.1, under 3 T magnetic field change) observed at Tc was attributed to the CO transition. The relatively high magnetic entropy change (-26.2 J/kg.K for x = 0 at 38 K, -6.5 J/kg.K for x = 0.1 at 83 K, under 5 T magnetic field change) observed at lower temperatures is attributed to very sharp step-like metamagnetic transitions. For further Sr concentrations, negative magnetic entropy change related to FM-PM phase transition was observed around Curie temperature. The peak temperature at which ΔS_M is seen has increased from 203 K for x = 0.18 to 267 K for x = 0.32 with the increasing Sr ratio. It is observed that the magnetic entropy change, under 1 T magnetic field change, decreases from -4.1 J/kg.K for x = 0.18 to -4.4 J/kg.K for x = 0.32. It was shown that $Pr_{0.67}Ca_{0.33}MnO_3$ compound also exhibits a CO-phase transition of around 200 K [32]. The more pronounced FM phase has observed below 56 K. The sharp step-like metamagnetic transitions observed at 5 K was attributed to phase separation. At 5 K, after the sample was exposed to 7 T magnetic fields, it was found that its magnetic properties show completely FM behavior. More interesting is that even when the sample is heated at a temperature above the CO transition, the sample remains in the FM phase without turning to its former magnetic properties. The change under 5 T magnetic field and the change in the highly negative magnetic entropy of 26.18 J/kg.K, observed at 38 K, are attributed to the metamagnetic transition. In recent years, Gencer [35] has examined the magnetic and magnetocaloric properties of $Pr_{0.68}Ca_{0.32-x}Bi_xMnO_3$ (x = 0, 0.1, 0.18, 0.26 and 0.32) compound. Only for the sample with x = 0, CO phase transition was observed at T_{co} = 200 K. For further Bi concentrations, the charge-order phase transition has disappeared. With increasing Bi concentration; the Curie temperature, saturation magnetism, and magnetic entropy change have been observed to decrease. The decrease in Curie temperature and saturation magnetization was attributed to the non-homogenous magnetic structure and the weakening of double-exchange interaction in Bi-doped samples. The decrease in the magnetic entropy change (1.094 J/kg.K for x = 0.1, 0.475 J/kg.K for x = 0.32, under 1 T magnetic field change) is due to the decrease in the saturation magnetism and the nature of phase transition. Magnetocaloric properties of several A-site doped manganites are listed in Table 3.

3.2.2 Mn-site substitution in manganites

As discussed above, in the case of A-site doping in manganites due to the different ionic radii and oxidation state of doping elements, the magnetic, transport and magnetocaloric properties change indirectly due to the change in the carrier density (Mn^{3+}/Mn^{4+}) and the structural parameters (Mn–O bond distance, Mn–O–Mn bond angle). In the case of Mn-site doping, in addition to structural parameters, since the Mn ions are replaced by different transition metals (TM), new exchange interactions (Mn–Mn, Mn–TM, TM–TM) occur between the newly added transition metal ions and Mn ions in the structure. This means that the magnetic, transport and and thus the magnetocaloric properties are directly affected by A-site doping. For this reason, Mn-site doping in manganite is quite attractive. Until today, many different transition metals (Fe, Cr, Cu, Al, Ni, Co, Sn, Si, Ru, Ti, Ga, V, Sb, Gd, In, Zn, Li) in FM metallic manganites ($La_{0.67}Ca_{0.33}Mn_{1-x}TM_xO_3$) and CO-insulator manganites ($La_{0.5}Ca_{0.5}Mn_{1-x}TM_xO_3$) have been replaced with Mn, and magnetic and magnetocaloric properties of these manganites have been investigated [157-213].

3.2.2.1 Mn-site substitution with Al

Tka [157] studied the effect of Al doping on the magnetic and magnetocaloric properties of $La_{0.57}Nd_{0.1}Sr_{0.33}Mn_{1-x}Al_xO_3$ ($0.0 \leq x \leq 0.3$) compounds. It has been shown that Curie temperature varies between 238 and 342 K and is closely related to Al concentration. It has been reported that magnetic entropy change increases from 2.31 J/kg.K for $x = 0$ to 3.58 J/kg.K for $x = 0.3$ under 1 T magnetic field change. Considerably large entropy change at a low magnetic field such as 1 T and in a large temperature range make these materials promising candidate in magnetic cooling applications. In another Al-doped $La_{0.7}Sr_{0.3}Mn_{1-x}Al_xO_3$ ($0 \leq x \leq 0.2$) compounds, it has been shown that these samples exhibit Griffiths like phase above the Curie temperature for $x \geq 0.15$[158]. This abnormal paramagnetic behavior is attributed to the presence of ferromagnetic clusters within paramagnetic domains. The Curie temperature has decreased from 366.74 K for $x = 0$ to 226.44 K for $x = 0.2$. It is reported that the magnetic entropy change, calculated using the phenomenological theoretical model under 0.01 T magnetic field decreased from $|\Delta S_M| = 162.34$ erg/g.K for $x = 0$ to $|\Delta S_M| = 9.022$ erg/g.K for $x = 0.2$. Dhahri [159] studied the effect of Al and Sn ions doping simultaneously on magnetic and magnetocaloric properties in the $La_{0.7}Ca_{0.1}Pb_{0.2}Mn_{1-x-y}Al_x Sn_yO_3$ ($0 \leq x, y \leq 0.075$) compounds. It has been observed that the Curie temperature decreased from 310 K for $x = 0$ to 290 K for $x = 0.075$. Magnetic entropy changes were reported as; $|\Delta S_M| = 3.7, 2.7, 2.3$ and 2 J/kg.K for $x,y = 0, 0.025, 0.05$ and 0.075, respectively.

3.2.2.2 Mn-site substitution with Co

In the study of Bau [160] for the $La_{0.7}Sr_{0.3}Mn_{0.05}Co_{0.95}O_3$ compound, which is rich in Co, Curie temperature was measured at 190 K. Magnetic entropy change under 4.5 T magnetic field change was reported as $|\Delta S_M| = 1.41$ J/kg.K. Although the $|\Delta S_M|$ value is small, the entropy change overspreading an extensive temperature range is a sought-after feature for the technological applications of magnetic cooling. In $La_{0.67}Pb_{0.33}Mn_{1-x}Co_xO_3$ compound, Curie temperature was measured as 297, 285, 272 and 260 K for $x = 0.15$, 0.2, 0.25 and 0.3, respectively, depending on the Co-doping [161]. Although Curie temperature decreases by Co content, magnetic entropy change is increasing. $|\Delta S_M|$ value was reported as 2.73, 2.92, 3 and 3.22 J/kg.K for $x = 0.15$, 0.2, 0.25 and 0.3 respectively under 1 T magnetic field change. In another study [162], Curie temperatures were reported as 360, 345, 324 and 316 K for $x = 0$, 0.03, 0.06 and 0.08 respectively depending on the Co concentration for $La_{0.67}Pb_{0.33}Mn_{1-x}Co_xO_3$ ($x = 0$, 0.03, 0.06, 0.08) compounds. In addition, magnetic entropy change was reported as 4.32, 0.58, 0.26, 0.19 J/kg.K. Although the two compounds are the same given in refrences [161] and [162], while the magnetic entropy change increases with Co concentration in the first compound while it decreases in the second compound. The discrepancies in these studies may be because of the sample preparation conditions or the Co ratios that are determined differently. Zhang [163] examined the magnetic and magnetocaloric properties of the $La_{0.7}Ca_{0.3}Mn_{1-x}Co_xO_3$ ($x = 0–0.05$) compounds. Magnetization measurements showed that Curie temperature decreased from 270 K for $x = 0$ to 215 K for $x = 0.05$, depending on Co concentration. Magnetic entropy change, under 1.5 T magnetic field change for $x = 0$ has been reported as 5.9 J/kg.K and for $x = 0.05$ as 4.8 J/kg.K. As can be seen from the results, the temperature range of semi-maximums (δT_{FWHM}) increases from 12 to 16 K, even though the $|\Delta S_M|$ change with increasing Co ratio drops a small amount. Despite the increase in δT_{FWHM} width, it has been determined that all samples still exhibit a phase transition from the first order. In $La_{0.7}Sr_{0.3}Mn_{1-x}Co_xO_3$ ($x = 0$, 0.05, 0.1) compounds, both Curie temperature and magnetic entropy change are strongly related to Co content [164]. For $x = 0$, 0.05 and 0.1, the Curie temperatures were measured as 338, 260 and 300 K respectively. Magnetic entropy change was reported as 1.36, 1.17 and 0.92 J/kg.K for $x = 0$, 0.05 and 0.1 respectively under 1.5 T magnetic field change. This decrease in $|\Delta S_M|$ value was attributed to the weakening of the DE interaction between Mn^{3+} and Mn^{4+} due to the increase Co contentration. In general, large saturation magnetization and the sudden change of magnetization in the phase transition region cause large magnetic entropy change. In the $La_{0.7}Sr_{0.3}Mn_{1-x}Co_xO_3$ alloy, replacement of Mn with Co ions reduces the number of ferromagnetically interacting $Mn^{3+}–Mn^{4+}$ pairs and leads to decrease in magnetization and $|\Delta S_M|$ value. Magnetization measurements for $La_{0.8}Ba_{0.1}Ca_{0.1}Mn_{1-x}Co_xO_3$ ($x = 0$, 0.05 and 0.10) compounds confirmed a large deviation between magnetization measured in zero field cooling (ZFC) conditions and magnetization curves measured in magnetic field cooling (FC) conditions [165]. This divergence between the ZFC and FC magnetization curves is connected to the canted ferromagnetic structure resulting from competition between FM and AFM interactions present in the structure. The Curie temperature has decreased from 282 K for x = 0 to 214 K for x = 0.1. In the study, it was found that all the samples exhibited a second-order phase transition. Magnetic entropy change under 5 T magnetic field change has been reported as 3.2, 2.5, and 0.8 J/kg.K, for x = 0, 0.05, and 0.1 respectively. In another study, the effect of Co-doping ($Pr_{0.7}Ca_{0.3}Mn_{1-x}Co_xO_3$, $0 \leq x \leq 0.1$) on magnetic and magnetocaloric properties of $Pr_{0.7}Ca_{0.3}MnO_3$ compound with CO content was examined [166]. While the $Pr_{0.7}Ca_{0.3}MnO_3$ compound exhibits CO configuration, it was observed that CO configuration disappeared in Co-doped compound and all samples exhibited a phase transition from a paramagnetic configuration to a ferromagnetic configuration. The Curie temperature has increased from 105 K for x = 0 to 116 K for x = 0.1. Magnetic entropy change under 5 T magnetic field change was reported as $|\Delta S_M|$ = 0.8, 2.2, 3.1 and 3.2 J/kg.K, for x = 0, 0.02, 0.05 and 0.1, respectively. As can be seen from the results, Co doping in FM ordered perovskite manganites have a negative effect on Curie temperature and magnetic entropy change, whereas in CO ordered manganites T_c and $|\Delta S_M|$ are observed to increase. Relative cooling power (RCP) was found to be 378.2 J/kg for $Pr_{0.7}Ca_{0.3}Mn_{0.95}Co_{0.05}O_3$ compound, under 5 T magnetic field change. This reported RCP value is 92 % of the RCP value measured in the same 5 T magnetic field of Gd element.

3.2.2.3 Mn-site substitution with Cr

In $La_{0.7}Sr_{0.3}Mn_{1-x}Cr_xO_3$ (x = 0, 0.2, 0.5 and 0.5) compounds, Curie temperature was measured as 369, 286, 242, 226 K for x = 0, 0.2, 0.4 and 0.5, respectively, depending on the Cr ratio [167]. A considerable decrease in Curie temperature was also observed in magnetic entropy change. For x = 0, 0.2, 0.4 and 0.5, $|\Delta S_M|$ value was reported as 1.27, 1.203, 0.473 and 0.279 J/kg.K. The first point that attracts attention here is that until x = 0.2, $|\Delta S_M|$ value has almost never changed. In higher Cr ratios, the $|\Delta S_M|$ value fairly reduced. The effect of Cr doping in $Pr_{0.6}A_{0.4}Mn_{1-x}Cr_xO_3$ (A = Ca and Sr, x = 0, 0.04) perovskite was examined [168]. In $Pr_{0.6}Ca_{0.4}Mn_{1-x}Cr_xO_3$ compound, the CO antiferromagnetic configuration observed for x = 0 was converted into a ferromagnetic metallic phase with the doping of Cr. In $Pr_{0.6}Sr_{0.4}Mn_{1-x}Cr_xO_3$ compound the structure for x = 0 is already in the FM metallic phase, and Curie temperature and saturation magnetization have been observed to decrease with the doping of Cr. In $Pr_{0.6}Ca_{0.4}Mn_{1-x}Cr_xO_3$ compound, magnetic entropy change was observed considerably small and negative value at 325 K for x = 0 sample. As we move towards low temperatures, ΔS_M increased and reached a maximum value at 255 K. At T_{CO} = 240 K , the sign of ΔS_M has changed. Around 230 K, the positive peak has observed (ΔS_M = 0.656 J/kg.K for ΔH = 5 T) . Under 200 K, ΔS_M become negative again. Unlike the sample of $Pr_{0.6}Ca_{0.4}MnO_3$, in the sample of $Pr_{0.6}Ca_{0.4}Mn_{0.96}Cr_{0.04}O_3$, ΔS_M has a negative value in a wide temperature range. ΔS_M measured as - 5.99 J/kg.K, under ΔH = 5 T and at T_c = 155 K. The results reveal that one of the ways to increase the magnetocaloric effect in CO manganites is to dope Cr to the structure. Abdelkhalek [169] reported that the compound of $La_{0.6}Sr_{0.4}Mn_{0.8}Fe_{0.1}Cr_{0.1}O_3$, which has been doped with Cr and Fe at the same ratio, exhibited a phase transition to second-order. When the transition temperature is around T_c = 212 K and under 1 and 5 T magnetic field changes, the $|\Delta S_M|$ measured as 0.43 and 1.75 J/kg.K, respectively. Oumezzine [170] studied the magnetic and magnetocaloric properties of $La_{0.67}Ba_{0.33}Mn_{0.9}Cr_{0.1}O_3$, compound. As a result of the calculations, it was understood that the sample exhibited a second-order magnetic phase transition. Under 5 T magnetic field change and at T_c = 324 K , magnetic entropy change $|\Delta S_M|$ = 4.2 J/kg.K and relative cooling power RCP = 238 J/kg.K values have made this compound one of the samples desirable in the magnetic cooling field at room temperature. Again, in another study, the effect of Cr doping on magnetic and magnetocaloric properties in a $La_{0.75}Sr_{0.25}Mn_{1-x}Cr_xO_3$ (x = 0.15, 0.20 and 0.25) compound was investigated systematically [171]. Curie temperatures were measured as T_c = 317, 278, and 253 K, for x = 0.15, 0.2 and 0.25, respectively. Magnetic entropy change under 5 T magnetic field change for x = 0.15, 0.2 and 0.25, were reported as respectively; $|\Delta S_M|$ = 3.5, 3.85 and 4.2 J/kg.K and relative cooling power RCP = 289, 323 and 386 J/kg. As T_c decreased due to Cr ratio, $|\Delta S_M|$ and RCP values were increased. Bellouz [172] studied the effect of Cr doping on the magnetic and magnetocaloric properties of $La_{0.65}Eu_{0.05}Sr_{0.3}Mn_{1-x}Cr_xO_3$ (x = 0.05, 0.1 and 0.15) compounds. Magnetization measurements explained that double-exchange interaction in the structure was weakened with increased Cr ratio. It has been observed that the Curie temperature decreases from 338 K for x = 0.05 to 278 K for x = 0.15, due to the weakening of the DE interaction. It has been determined that the structure exhibits a second-order magnetic phase transition. It was observed that the magnetic entropy change decreased from 4.04 J/kg.K for x = 0.05 to 0.78 J/kg.K for x = 0.15, under 5 T magnetic field change. The RCP value reported for $La_{0.65}Eu_{0.05}Sr_{0.3}Mn_{1-x}Cr_xO_3$ compound is about 54% of the RCP value of pure Gd element, making these materials one of the most striking examples of magnetic cooling. The Rietveld analysis of $La_{0.7}Sr_{0.1}Ca_{0.2}Mn_{1-x}Cr_xO_3$ (x = 0, 0.05 and 0.1) compounds has revealed that Cr doping changes structural parameters such as Mn-O bond length, Mn–O–Mn bond angle [173] etc. It is observed that the replacement of Mn ions with Cr ions reduces the 2p-3d hybridization between O and Mn and therefore the bandwidth of Mn ions. With increasing Cr ratio, Curie temperature decreases from 294 K to 255 K, confirming these results. Under 5 T magnetic field change, the magnetic entropy change decreased from 6.2 J/kg.K for x = 0 to 3.8 J/kg.K for x = 0.1. At the same time, the relative cooling power, RCP, has increased from 234.5 J/kg for x = 0 to 240 J/kg for x = 0.1. These results show that Cr doping in some perovskites instead of Mn restrict the improvement of magnetocaloric properties. This situation is explained by the weakening of the Mn^{3+}–Mn^{4+} ferromagnetic interaction in Cr-doped manganites. In $La_{0.5}Sr_{0.5}Mn_{1-x}Cr_xO_3$ (x = 0.05, 0.1, 0.15 and 0.2) compounds, depending on the Cr doping, the Curie temperature was observed to decrease from 319 K for x = 0.05 to 251 K for x = 0.2 [174]. From Arrott plots, it was observed that the magnetic phase transition is second-degree. Magnetic entropy change has been reported as 2.77, 1.91, 1.59 and 1.35 J/kg.K, for x = 0.05, 0.1, 0.15 and 0.2, respectively, under 5 T magnetic field change. Under 5 T magnetic field change, cooling power for x = 0.05, 0.1, 0.15 and 0.2 was reported as RCP = 288, 213, 152 and 122 J/kg. Gencer [175] examined the effect of Cr doping on the magnetic and magnetocaloric properties of $La_{0.94}Bi_{0.06}Mn_{1-x}Cr_xO_3$ (x = 0, 0.05, 0.1, 0.15, 0.2 and 0.25) compounds. It has been observed that all samples have orthorhombic symmetry. In this study, we have determined that the Curie temperature decreases from 209 K for x = 0 to 127 K for x = 0.25, with a Cr concentration, and the magnetization measurements revealed that the saturation magnetization (M_s) decreases from 86.13 emu/g for x = 0 to 35.69 emu/g for x = 0.25. The decrease in Curie temperature and saturation magnetization in Cr was attributed to the weakening of ferromagnetic interactions in Cr-doped samples. Magnetic entropy change and relative cooling power were reported as $|\Delta S_m|$ = 5.51, 2.42, 2.47, 1.99, 1.61, 1.26 J/kg.K and RCP = 264.53, 253.15, 202.21, 162.46, 129.86, 100.41 J/kg for x = 0.0, 0.05, 0.1, 0.15, 0.2 and 0.25, respectively, under 5 T magnetic field change. The decrease in the magnetocaloric effect was attributed to the reduction in the saturation magnetism by the doping of Cr.

Studies have revealed that the replacement of Cr with Mn ions in manganites often reduces the magnetocaloric properties and Curie temperature [167-175]. In most studies, the reduction of T_c was explained by the antiferromagnetic superexchange interactions between Cr^{3+}–O–Mn^{3+} and Cr^{3+}–O–Cr^{3+}. Magnetization measurements have proven strong competition between FM double-exchange interaction and AFM super-exchange interaction in Cr-doped compounds. Because of the smaller ionic radius of Cr^{3+} (0.615 Å) compared with Mn^{3+} (0.645 Å), a structural effect is expected in Cr doped manganites. In this case, structural deformations may have an indirect effect on the ferromagnetic DE interaction.

3.2.2.4 Mn-site substitution with Fe

Although the effect of Fe deposition instead of Mn ions on the magnetic and conductivity properties of manganites has been studied for many years, the effect on magnetocaloric properties has been studied in recent years and quite interesting results have been reported [176-183,51]. Nisha [176] investigated the structural magnetic and magnetocaloric properties of nanocrystal $La_{0.67}Ca_{0.33}Mn_{1-x}Fe_xO_3$ (x = 0.05, 0.2) compounds. The average particle size was reported as 15 nm and 42 nm respectively for x = 0.05 and 0.2. Although the ionic radius (0.645 Å) of the Fe^{3+} ion is equal to the Mn^{3+} ion, the increase of the lattice parameters and unit cell volumes due to the Fe ratio can be caused by the lattice distortions resulting from the random distribution of Fe and Mn ions. The other possibility is that there are Fe^{3+} ions in the structure as well as Fe^{4+} ions, because Fe^{4+} ion has a higher ionic radius (0.585 Å) than Mn^{4+} (0.53 Å) ion. Magnetization measurements have revealed that $La_{0.67}Ca_{0.33}Mn_{0.95}Fe_{0.05}O_3$ compound has a super paramagnetic configuration and $La_{0.67}Ca_{0.33}Mn_{0.8}Fe_{0.2}O_3$ compound has spin-glass type configuration. In many studies, although the $La_{0.67}Ca_{0.33}MnO_3$ alloy has been proven to show a first-order phase transition, the Arrott drawings showed that both of the Fe doped samples exhibited a second-order magnetic phase transition. The Curie temperature decreased from 162 K for x = 0.05 to 92 K for x = 0.2. $|\varDelta S_M|$ decreased from 2.3 J/kg.K for x = 0.05 to 0.3 J/kg.K for x = 0.2. Magnetization measurements have explained that both samples still do not reach saturation under the 5 T magnetic field. Such a small magnetic entropy change was attributed to the second-order magnetic phase transition and the samples do not reach saturation even if 5 T magnetic field applied. As distinct from, the effect of Fe doping on magnetocaloric properties of antiferromagnetic insulating $LaMnO_3$ compound was investigated [177]. It has been reported that $LaMn_{0.9}Fe_{0.1}O_3$ compound shows a second-order magnetic phase transition. Magnetization measurements have shown that the structure has a short-range ferromagnetic configuration. It has been reported that the magnetic entropy change at 137 K was $|\varDelta S_M|$ = 3.8 J/kg.K, under 5 T magnetic field change. X-ray studies for $Nd_{0.67}Ba_{0.33}Mn_{1-x}Fe_xO_3$ ($0 \leq x \leq 0.1$) compounds have showed that unlike in nano-sized manganites [176], Fe^{3+} and Mn^{3+} ions in the polycrystalline structure have no significant influence on the structural parameters of the compound because they have the same ionic radius [178]. Magnetization measurements have showed that the compound exhibits ferromagnetic behavior for x = 0 and 0.02, and spin-glass-like behavior for $x \geq 0.05$. The Curie temperature was measured as 150, 131, 61, 50 and 40 K for x = 0, 0.02, 0.05, 0.007 and 0.1 respectively. Magnetic entropy change has been reported as $|\varDelta S_M|$ = 3.91 and 2.97 J/kg.K for x = 0 and 0.02, under 5 T magnetic field change. The relative cooling power was reported as RCP = 265 and 242 J/kg. Magnetization measurements for $La_{0.67}Sr_{0.22}Ba_{0.11}Mn_{1-x}Fe_xO_3$ ($0 \leq x \leq 0.3$) compounds have showed that the samples for x = 0 and 0.1 exhibit FM-PM phase transition [179]. A large divergence observed between ZFC and FC curves for $0.2 \leq x \leq 0.3$ samples is a clear evidence of competition between FM and AFM interactions. Curie temperature decreased from 360 K for x = 0 to 94 K for x = 0.2. $|\Delta S_M|$ and RCP were reported as 2.46, 2.43, 0.91 J/kg.K and 169, 241, 70 J/kg for x = 0, 0.1 and 0.2 respectively under 5 T magnetic field change. In $La_{0.8}Ca_{0.2}Mn_{1-x}Fe_xO_3$ (x = 0, 0.01, 0.15, 0.2) compounds, depending on the Fe ratio, Curie temperature decreased from 223 K for x = 0 to 70 K for x = 0.2 [180]. Magnetic entropy change for x = 0, 0.01, 0.15 and 0.2, were reported to be 4.42, 4.32, 1.6 and 0.54 J/kg.K respectively under 5 T magnetic field change. The results confirmed that magnetic properties are strongly related to Fe concentration. In 2016, the effect of Fe doping on the magnetic and magnetocaloric properties of $La_{0.67}Sr_{0.33}Mn_{1-x}Fe_xO_3$ (x = 0, 0.05, 0.1, 0.2) [181] and $La_{2/3}Ba_{1/3}Mn_{1-x}Fe_xO_3$ (x = 0.0–0.10) [182] compounds, which exhibit similar magnetic properties, was examined. In $La_{0.67}Sr_{0.33}Mn_{1-x}Fe_xO_3$ compound, Curie temperature decreased from 355 K to 100 K, depending on Fe concentration. Similar behavior at Curie temperature has been observed in $La_{2/3}Ba_{1/3}Mn_{1-x}Fe_xO_3$ compound. In $La_{0.67}Sr_{0.33}Mn_{1-x}Fe_xO_3$ compound, magnetic entropy change determined as 1.66, 0.59, 0.79 and 0.42 J/kg.K for x = 0, 0.05, 0.1 and 0.2 respectively under 3 T magnetic field change, In $La_{2/3}Ba_{1/3}Mn_{1-x}Fe_xO_3$ compound, the magnetic entropy change first increased from 1.06 J/kg.K for x = 0 to 1.46 J/kg.K for x = 0.025. For the further Fe concentrations ($x \geq 0.05$), $|\Delta S_M|$ decreased to 1.14 J/kg.K. under 2.5 T magnetic field change In $La_{0.7}Te_{0.3}Mn_{1-x}Fe_xO_3$ (x = 0.1 and 0.3) compound, it was seen that the compound was a rhombohedral structure for x = 0.1 and orthorhombic structure for x = 0.3 [183]. The Curie temperature has decreased from 171 K for x = 0.1 to 78 K for x = 0.3, $|\Delta S_M|$ and RCP were reported as 1.17 J/kg.K and 80 J/kg for x = 0.1 and 0.44 J/kg.K and 49 J/kg for x = 0.3 under 2 T magnetic field change. Gencer [51] examined the magnetic and magnetocaloric properties of $La_{0.94}Bi_{0.06}Mn_{1-x}Fe_xO_3$ (x = 0, 0.05, 0.075 and 0.1) compounds. It has been observed that saturation magnetism, Curie temperature, and magnetic entropy change have decreased with increasin Fe content. The Curie temperature was measured as 209, 185, 160, 156 K for x = 0, 0.05, 0.075, 0.1, respectively. Magnetic entropy change and RC were reported as 5.51 J/kg.K and 259 J/kg for x = 0, and 2.89 J/kg.K and 225 J/kg for x = 0.075 under 5 T magnetic field change. The decrease in the magnetic entropy change is attributed to the decrease in the saturation magnetism and the change in the nature of the phase transition from first to second-order.

In general, substitution of Mn^{3+} with Fe^{3+} in Fe-doped manganites resulted in a decrease in Curie temperature [176-183, 51]. Since the ionic radius (0.645 Å) of the Fe^{3+} ion is equal to the Mn^{3+} ion, it is not expected any change in structural parameters. In spite of this, significant changes in lattice parameters and unit cell volume were observed due to Fe concentrations in some manganites. [176]. This situation is attributed to structural distortion due to the random distribution of Fe and Mn ions in crystal structure. Another possibility of these structural distortion is the presence of Fe^{4+} ions along with Fe^{3+} ions in crystal structure. Because Fe^{4+} (= 0.585 Å) ions have larger ionic radius than Mn^{4+} (0.53 Å) ions. However, the effect of structural parameters on magnetic properties is considerably small. The decrease of T_c in the Fe doped manganites is mainly due to the formation of new antiferromagnetic interactions by the addition of Fe^{3+} ions. In many studies, it has been concluded that Fe^{3+}–O–Fe^{3+} and Fe^{3+}–O–Mn^{3+} antiferromagnetic super exchange interactions induced by Fe doping weakens the ferromagnetic Mn^{3+}–O–Mn^{4+} double-exchange interactions and consequently the Curie temperature decreased.

3.2.2.5 Mn-site substitution with Cu

In $La_{0.77}Sr_{0.23}Mn_{1-x}Cu_xO_3$ (0.1 $\leq x \leq$ 0.3) compounds, it has been observed that Curie temperature decreases from 325 K for x = 0.1 to 242 K for x = 0.3 with increasing Cu content [184]. Interestingly, magnetic entropy change was decreased with increasing Cu content in the samples have rohmboherderal phase till $x \leq 0.2$. However, for $x \geq 0.3$, the structure has transformed to the orthorhombic phase, and the magnetic entropy change has begun to increase again with increasing Cu ratio. $|\Delta S_m|$ and RCP were repoerted as 4.41, 2.68, 3.36 J/kg.K and 570, 396, 330 J/K for x = 0.1, 0.2 and 0.3 respectively, under 1 T magnetic field change. RCP values obtained for $La_{0.77}Sr_{0.23}Mn_{1-x}Cu_xO_3$ compounds were claimed to be around 60% of RCP values reported for pure Gd. Moreover, it is claimed that observed magnetic entropy change value of 4.41 J/kg.K for x = 0.1 under 1 T magnetic field is 26 % higher than that of pure Gd. For this reason, it has been stated that Cu-doped manganites may have potential refrigerant for magnetic cooling around room temperature due to their high $|\Delta S_m|$ and RCP values. In addition to the study of Cu doping in LaSrMnO$_3$ manganites, the effect of Cu doping on the magnetic and magnetocaloric properties of $La_{0.7}Sr_{0.25}Na_{0.05}MnO_3$ [185] and $La_{0.65}Sr_{0.3}Ce_{0.05}MnO_3$ [186] compounds were investigated. $La_{0.7}Sr_{0.25}Na_{0.05}Mn_{1-x}Cu_xO_3$ (x = 0, 0.05, 0.10, 0.15, 0.20) compound [185] is rhombohedral in all Cu proportions. X-ray photoelectron spectroscopy (XPS) revealed that Cu^{2+} and Cu^{3+} ions coexist in x = 0.15 and 0.2 samples. Magnetization measurements showed that Curie temperature and magnetic entropy change decreased with the increasing Cu ratio. Curie temperatures were reported as 362, 359, 320, 274, 167 K for x = 0, 0.05, 0.1, 0.15 and 0.2, respectively. Magnetic entropy changes were reported as 2.2, 2.02, 1.75, 1.39, 0.67 J/kg.K for x = 0, 0.05, 0.1, 0.15 and 0.2 under 2 T magnetic field change. Although T$_c$ and $|\Delta S_M|$ values decrease with Cu ratio, relative cooling power has an average value of around 87 J/kg. Similarly, $La_{0.65}Sr_{0.3}Ce_{0.05}Mn_{1-x}Cu_xO_3$ (0 $\leq x \leq$ 0.15) [186] compounds were found to be in the rhombohedral structure for all Cu concentrations. From the analysis of the crystallographic data, it was observed that there was a strong correlation between structural characteristics and magnetic properties. For example, the decrease in Curie temperature is thought to be related to the deformation of MnO_6 octahedrons in Cu-doped samples. Experimental measurements have confirmed that the Cu doping in the structure instead of Mn ions disrupts the formation of the Mn^{3+}–O–Mn^{4+} bond and weakens the ferromagnetic DE interaction between Mn^{3+} and Mn^{4+} ions. The Curie temperature is measured as 360, 330, 305, 275 K for $x = 0, 0.05, 0.1, 0.15$, respectively. Magnetic entropy change has been reported as 1.49, 1.34, 1.5, 1.08 J/kg.K for $x = 0, 0.05, 0.1, 0.15$, respectively under 1 T magnetic field change. The variation of $|\Delta S_M|$ values with Cu content is similar to the results obtained by Hagary [184]. In $La_{0.77}Sr_{0.23}Mn_{1-x}Cu_xO_3$ ($0.1 \leq x \leq 0.3$) compounds, while $|\Delta S_M|$ decreases for $x \leq 0.2$, it begins to increase due to the structural phase transition for $x \geq 0.3$. Interestingly, in the $La_{0.65}Sr_{0.3}Ce_{0.05}Mn_{1-x}Cu_xO_3$ ($0 \leq x \leq 0.15$) compound, the $|\Delta S_M|$ value decreases until $x = 0.05$ and increases again until $x = 0.15$. Nanto [187] studied the magnetic and magnetocaloric properties of $La_{0.7}Ca_{0.3}Mn_{1-x}Cu_xO_3$ ($0.0 \leq x \leq 0.03$) compounds. The Arrott plots show that all Cu-doped samples exhibit a first-order phase transition. The Curie temperatures were determined to be 260, 248, 230, 217 K and $|\Delta S_M|$ and RCP values were reported as 4.32, 3.46, 2.98, 2.74 J/kg.K and 45, 42, 39, 47 J/kg for $x = 0, 0.01, 0.02$ and 0.03 respectively at 1 T for magnetic field change

3.2.2.6 Mn-site substitution with Ni

In 2012, Zhang [188] examined the effect of Ni doping on the magnetic and magnetocaloric properties in $La_{0.7}Sr_{0.3}Mn_{1-x}Ni_xO_3$ ($x = 0, 0.01, 0.02, 0.03$) compounds. In compound, Ni ions were shown to be in the Ni^{2+} valence state. In this case, increasing Ni^{2+} content in the structure causes the increase of Mn^{4+} concentration. As a consequence, the increase in the number of Mn^{4+} ions causes a decrease ferromagnetically interacting $Mn^{3+}-O-Mn^{4+}$ pairs. Finally, it causes a considerable decrease in ferromagnetic interactions and therefore Curie temperature decreases in Ni-doped compounds. Curie temperatures were determined to be 362, 356, 350 and 347 K for $x = 0, 0.01, 0.02$ and 0.03 respectively. Magnetic entropy change has been reported as 2.33, 2.27, 2.26 and 2.21 J/kg.K, for $x = 0, 0.01, 0.02$ and 0.03 respectively, under 1.5 T magnetic field change. As can be seen from the results, in the $La_{0.7}Sr_{0.3}MnO_3$ compound, while a small amount of Ni doping has a insignificant effect on magnetic entropy change, the temperatures at which maximum entropy change occurs has shifted towars to room temperature. Semli [189] examined the magnetic and magnetocaloric properties in $Pr_{0.7}Ca_{0.3}Mn_{1-y}Ni_yO_3$ ($0 \leq y \leq 0.1$) compounds. In $y = 0$ sample, the observed small maximum in magnetization curve at 215 K was interpreted as existence of CO phase. In the magnetization curves obtained for further Ni concentrations, the small maximum corresponds to CO was observed to disappear. As a result, with the Ni doping it was observed to eliminate CO phase observed for $y = 0$ sample. Also, depending on the Ni concentration, Curie temperature decreased from 106 K for $y = 0.02$ to $T_c = 118.4$ K for $y = 0.1$. Magnetic entropy change for $y = 0.02, 0.05$ and 0.1 has been reported as 2, 2.96 and 2.94 J/kg.K and RCP = 239.5, 352.2 and 308.7 J/kg, under 5 T magnetic field change. Oumezzine [190] studied the magnetic and magnetocaloric properties in $La_{0.6}Pr_{0.1}Ba_{0.3}Mn_{1-x}Ni_xO_3$ ($0 \leq x \leq 0.3$) nanocrystal compound. With increasing Ni concentration in the structure results an increase in the rate of Mn^{4+} ions compared to Mn^{3+} ions. In other words, this leads to an increase in the hole concentration and therefore to a decrease in density of e_g electrons. This situation caused the magnetization and Curie temperature in the system to decrease from 215 K to $x = 0$ to 131 K for $x = 0$. The magnetic entropy change has decreased from 1.97 J/kg.K for $x = 0$ to 0.65 J/kg.K for $x = 0.3$, under 5 T magnetic field change. Similarly, the relative cooling power decreased from RCP = 230 J/kg for $x = 0$ to 62 J/kg for $x = 0.3$, depending on Ni concentration. In $La_{0.7}Ca_{0.3}Mn_{1-x}Ni_xO_3$ ($x = 0$, 0.02, 0.07 and 0.1) nanocrystal compound, similar to previous Ni-doped manganites, it has been observed that the Curie temperature decreased from 264 K for $x = 0$ to 174 K for $x = 0.1$ [191]. The decrease in Curie temperature with increasing Ni concentration is explained by the weakening of the FM double-exchange interaction between $Mn^{3+}–O–Mn^{4+}$. The Arrott plots determine that all samples exhibit a second-order magnetic phase transition. Interestingly, the second-order phase transition of the $La_{0.7}Ca_{0.3}MnO_3$ compound is attributed to the nanosize of the samples. The magnetic entropy change was reported as 0.85, 0.77, 0.63 and 0.59 J/kg.K for $x = 0$, 0.02, 0.07 and 0.1 respectively under 1.5 T magnetic field change.

As shown in the studies on Ni doping in manganites, Ni^{2+} substitution for Mn^{3+} suppress the FM double-exchange interaction and consequently results in the decrease in magnetization, Curie temperature and magnetic entropy in Ni-doped manganites. The increase in number of Ni^{2+} causes a decrease in the number of Mn^{3+} ions (e_g electron density) and therefore the increase in the number of Mn^{4+} ions (hole concentration). This situation results in a decrease in the number of $Mn^{3+}–O–Mn^{4+}$ pairs interacting ferromagnetically in the structure, and an increase in the number of $Ni^{2+}–O–Ni^{2+}$ and $Mn^{4+}–O–Mn^{4+}$ pairs interacting antiferromagnetically.

3.2.2.7 Mn-site substitution with Ga

In $La_{0.7}Ca_{0.15}Sr_{0.15}Mn_{1-x}Ga_xO_3$ ($x = 0$, 0.025, 0.05, 0.075 and 0.1) compound, it was reported that the Curie temperature decreased from 336.5 K for $x = 0$ to 244.5 K for $x = 0.1$, depending on the Ga ratio [192]. The Arrott plots revealed that the samples exhibited first-order phase transition up to $x = 0.05$ ratio and for the further Ga concentrations, the nature of phase transition transformed from first-order to second-order. Magnetic entropy change decreased from 5.15 J/kg.K for $x = 0$ to 1.86 J/kg.K for $x = 0.1$ with increased Ga concentration under 5 T magnetic field change. Again, magnetocaloric properties have been examined in similar $La_{0.75}Ca_{0.08}Sr_{0.17}Mn_{1-x}Ga_xO_3$ ($0 \leq x \leq 0.2$) compounds [193]. Magnetization measurements showed that the Curie temperature decreases with the increasing Ga ratio. Curie temperatures were reported as 336, 285, 241 and 135 K for x = 0, 0.05, 0.1 and 0.2 respectively. Magnetic entropy change were reported as 2.87, 1.92, 1.57, 1.17 J/kg.K and RCP = 97.5, 83, 101, 89 J/kg for x = 0, 0.05, 0.1, 0.2 respectively under 2 T magnetic field change. The effect of non-magnetic Ga-ion on magnetic and magnetocaloric properties of $La_{0.7}(Ba, Sr)_{0.3}Mn_{1-x}Ga_xO_3$ (x = 0, 0,1, 0,2) compounds was examined by Tlili [194]. Magnetization measurements show that Curie temperature decreased from 316 K for x = 0 to 300 K for x = 0.2. The Arrott plots reveal that the samples exhibit a second-order phase transition. Magnetic entropy change were reported as 1.27, 1.16, 1.02 J/kg.K and RCP = 75.74, 72.35, 71.29 J/kg for x = 0, 0.1 and 0.3 respectively under 2 T magnetic field change. When Mn ions are replaced with non-magnetic Ga^{3+} ions, the number of Mn^{3+} ions decreases and some $Mn^{3+}-O-Mn^{4+}$ bonds existing in the structure transform into the form of $Ga^{3+}-O-Mn^{4+}$ and $Ga^{3+}-O-Ga^{4+}$. Also, it causes a decrease in the mobility of e_g electrons. Thus, the doping of Ga^{3+} ions in manganites causes FM double-exchange interaction to weakens and consequently magnetization and Curie temperature to decrease.

3.2.2.8 Mn-site substitution with Ti

Kallel [195] calculated the magnetic entropy change of $La_{0.70}Sr_{0.30}Mn_{0.90}Ti_{0.10}O_3$ as a function of Ti concentration, by usin magnetization measurements and Landau theory. The Curie temperature was repoerted as 369 K for $La_{0.70}Sr_{0.30}MnO_3$ compound and 210 K for $La_{0.70}Sr_{0.30}Mn_{0.90}Ti_{0.10}O_3$ compound respectively. Magnetic entropy change and RCP values were determined to increase from 2.31 J/kg.K and 69 J/kg for x = 0 to 2.94 J/kg.K and 288 J/kg for x = 0.1 under 5 T magnetic field change. The RCP value obtained for $La_{0.70}Sr_{0.30}Mn_{0.90}Ti_{0.10}O_3$ is 70% of the value of Gd element's RCP value measured in the same field and moreover greater than that of $La_{0.70}Sr_{0.30}MnO_3$ alloy where many different TM elements are doped. In general, the compliance between experimental and theoretical calculations shows the effect of magneto elastic interaction and electron interaction on the magnetocaloric properties. The change of the Exchange interaction energy by the applied magnetic field around the phase transition adds an additional contribution to the magnetic entropy change. Phong [196] systematically examined the effect of Ti doping on magnetic and magnetocaloric properties in $La_{0.7}Sr_{0.3}Mn_{1-x}Ti_xO_3$ ($0 \leq x \leq 0.3$) compounds. Magnetization studies have revealed that doping of Ti weakens the FM double-exchange interaction, and causes to decrease the Curie temperature (364, 307, 236, 132 and 55 K, for x = 0, 0.05, 0.1, 0.2 and 0.3). Besides, by increasing the Ti ratio, the temperature of the metal-insulator transition decreases and transforms into the insulator phase for $x \geq 0.2$ samples. Magnetic entropy change under 0.01 T magnetic field change has been reported as $|\Delta S_M|$ = 0.0162, 0.0042 and 0.0071 J/kg.K, for x = 0, 0.05 and 0.1, using the phenomenological model of Hamadin [75]. The behavior of the results is consistent with the results given in reference [40]. Again, it was observed that Curie temperature decreased from 363 K for $x = 0$ to 125 K for x = 0.2, in Ti-doped $La_{0.7}Sr_{0.25}Na_{0.05}Mn_{1-x}Ti_xO_3$ ($0 \leq x \leq 0.2$) compounds [197]. Magnetic entropy change and RCP values were reported to decrese from 4.34 J/kg.K and 298 J/kg for $x = 0.2$ to 2.03 J/kg.K and 273 J/kg for $x = 0.2$, under 5 T magnetic field change. In the same year, the effect of Ti doping on magnetocaloric properties in $La_{0.67}Ba_{0.22}Sr_{0.11}Mn_{1-x}Ti_xO_3$ ($x = 0$, 0.1, 0.2 and 0.3) compound was examined [198]. Similarly, the Curie temperature has decreased from 344 K for $x = 0$ to 267 K for $x = 0.1$. $|\varDelta S_M|$ and RCP were reported to be 2.75 J/kg.K and 290 J/kg for $x = 0$ and 1.33 J/kg.K and 255 J/kg for $x = 0.1$, under 5 T magnetic field. Smiy [199] examined the magnetic and magnetocaloric properties of $La_{0.5}Pr_{0.2}Sr_{0.3}Mn_{1-x}Ti_xO_3$ ($x = 0.0$ and 0.1) compounds. The Curie temperature has decreased from 280 K for $x = 0$ to 123 K for $x = 0.1$. $|\varDelta S_m| = 1.969$ J/kg.K and RCP = 285 J/kg for $x = 0$ decreased to $|\Delta S_M| = 1.309$ J/kg.K and RCP = 162 J/kg for $x = 0.1$, under 5 T magnetic field change.

If we look at the results in general, it is observed that Ti doping has a negative effect on Curie temperature and magnetic entropy change in manganites. One of the highlights is that even small amounts of Ti doping have a substantial effect on magnetic and magnetocaloric properties compared to other Mn site dopings. As can be seen from the results, the Ti doping in minimal ratios has caused a significant reduction in Curie temperature. Since the Ti^{4+} ions do not have any 3d electrons, they do not exhibit any magnetic properties. When Mn ions are substituted with Ti ions, the ferromagnetic $Mn^{3+}-O-Mn^{4+}$ interactions cause a reduction. Since Ti^{4+} ions do not exhibit magnetic properties and there are no further interactions to suffice this decline, the decrease in T_c is faster than in other transition metals.

3.2.2.9 Mn-site substitution with V

Kolat [25] examined the magnetic, electrical and magnetocaloric properties of $La_{0.67}Ca_{0.33}Mn_{0.9}V_{0.1}O_3$ compound. In the study, V atoms in the structure were shown to be in the state of V^{4+}. For $La_{0.67}Ca_{0.33}MnO_3$, compound, only one magnetic phase transition was observed at $T_C = 267$ K, whereas two different phase transitions were observed for $La_{0.67}Ca_{0.33}Mn_{0.9}V_{0.1}O_3$ compound at $T_{c1} = 223$ K and $T_{c2} = 190$ K. Two different magnetic transitions observed in the magnetization curve of V-doped compound were attributed to the presence of two ferromagnetic phases in the structure. From EDX analysis, it was concluded that two different transition temperatures belong to the $La_{0.67}Ca_{0.33}Mn_{0.9}V_{0.1}O_3$ and $La_{0.4}Ca_{0.6}Mn_{0.21}V_{0.79}O_3$ phases. The magnetization curves, under 3 T magnetic field and at 5 K, showed that the saturation magnetization decreased from 92 emu/g for $x = 0$ to 77 emu/g for $x = 0.1$. The reduction of the saturation magnetism in V-doped compound means that V in the structure suppresses the ferromagnetic interactions. In V-doped compound, the decrease of the Curie temperature is one of the evidence of the weakening of ferromagnetism. In the magnetic entropy change curve obtained for $La_{0.67}Ca_{0.33}Mn_{0.9}V_{0.1}O_3$ alloy, two peaks were observed around T_{c1} and T_{c2} temperatures. Under 1 T magnetic field change, the maximum magnetic entropy change obtained for $La_{0.67}Ca_{0.33}MnO_3$ has a value of 4 J/kg.K, whereas the value for V-doped $La_{0.67}Ca_{0.33}Mn_{0.9}V_{0.1}O_3$ compound has been reported as $|\Delta S_M|$ = 2.4 J/kg.K. As it can be seen, in V-doped compound, $|\Delta S_M|$ value has decreased significantly. This decrease in magnetic entropy change is explained by the weakening of ferromagnetism and therefore the decrease in saturation magnetism, in V-doped compound. In another V-doped $La_{0.6}Nd_{0.1}(CaSr)_{0.3}Mn_{0.9}V_{0.1}O_3$ compound, magnetization measurements have revealed that the compound exhibits a second-order magnetic phase transition [200]. Curie temperature measured as 298 K. Magnetic entropy change and RCP values were reported as 4.266 J/kg.K and 205.35 J/kg respectively under 5 T magnetic field change. The obtained $|\Delta S_M|$ value is almost two times higher than the same V-doped $La_{0.67}Ca_{0.33}Mn_{0.9}V_{0.1}O_3$ sample. Also, the temperature (298 K) at which $|\Delta S_M|$ change is observed in this example corresponds to room temperature. The RCP value obtained for this sample is about 49.72 % of the value obtained for pure Gd at the same magnetic field. Therefore, this compound is one of the desired examples for magnetic cooling at room temperature. Mansouri [201] studied the magnetic and magnetocaloric properties of $La_{0.7}Sr_{0.2}Ca_{0.05}Li_{0.05}Mn_{1-x}V_xO_3$ (x = 0 and x = 0.05) compounds. The Curie temperature has decreased from 271 K for x = 0 to 266 K for x = 0.05. It has been determined that all samples exhibit a second-order phase transition. $|\Delta S_M|$ and RCP values were reported to be 5.4 J/kg.K and 211.5 J/kg for x = 0, and 4.8 J/kg.K and 195.5 J/kg for x = 0.05 at 5 T magnetic field. Similarly, it has been reported that Curie temperature decreased from 262 K for x = 0 to 208 K for x = 0.5, depending on V ratio in the $La_{0.65}Ca_{0.35}Mn_{1-x}V_xO_3$ ($0 \leq x \leq 0.5$) compound [202]. The magnetic entropy changes under 5 T magnetic field change were reported as 5.5 J/kg.K for x = 0, 3.36 J/kg.K for x = 0.1, and 5.25 J/kg.K for x = 0.5. As pointed in the results, the value of $|\Delta S_M|$ decreased until it reached x = 0.1 and then increased again with the increasing V ratio. The relative cooling power increased from the 125 J/kg for x = 0.1 to 207 J/kg for x = 0.5, with the increasing V ratio.

3.2.2.10 Mn-site substitution with Sn

Dhahri [203] examined the effect of Sn doping on the magnetic and magnetocaloric properties of the $La_{0.67}Ba_{0.33}Mn_{1-x}Sn_xO_3$ (x = 0.05, 0.1 and 0.15) compounds. Curie temperature was found to be 340, 325 and 288 K for x = 0.05, 0.1 and 0.15 respectively. As can be seen clearly, the Curie temperatures of these Sn-doped manganites contain room temperature. $|\Delta S_M|$ and RCP values were reported as 1.9 J/kg.K and 101 J/kg for x =

0.05, 2.27 J/kg.K and 120 J/kg for $x = 0.1$ and 2.49 J/kg.K and 123 J/kg for $x = 0.15$ under 2 T magnetic field change. As can be seen from the results, while Curie temperature decreases with Sn ratio, $|\Delta S_M|$ and RCP values increase. Sn is a non-magnetic cation with $4d^{10}5s^{0}5p^{0}$ electron configuration. In this case, the doping of Sn cannot be expected to have a direct magnetic contribution to the system. As can be seen from the experimental results, the doping of Sn^{4+} in $La_{0.67}Ba_{0.33}Mn_{1-x}Sn_xO_3$ compound has revealed that magnetic properties have changed considerably. Given the charge neutrality, the addition of Sn^{4+} to the structure causes the mean valence states of Mn atoms to convert to the Mn^{3+} state. In this case, the decrease in the number of Mn^{4+} ions causes a decrease in e_g electron density. In addition, replacement of Mn^{4+} (0.53 Å) with Sn^{4+} (0.69 Å) ions which have larger ionic radius leads to changes in the structural parameters such as Mn–O bond length and Mn–O–Mn bond angle. This causes to weaken the FM double-exchange interaction in the Sn doped structure. As a result of the increase in the ratio of Sn in $La_{0.57}Nd_{0.1}Sr_{0.33}Mn_{1-x}Sn_xO_3$ $(0.05 \leq x \leq 0.30)$ compounds, the Curie temperature (T_c = 282, 224, 187 and 158 K for $x = 0.05, 0.1, 0.15$ and 0.2, respectively) decreased [204]. In the examinations, it was found that the samples showed a second-order magnetic phase transition. Magnetic entropy change has been reported as $|\Delta S_M|$ =2.8 J/kg.K for $x = 0.0.5$ and 3.22 J/kg.K for $x = 0.1$, under 5 T magnetic field change. A very small amount of Sn doping in $La_{0.7}Ca_{0.3}Mn_{1-x}Sn_xO_3$ $(x = 0.0, 0.02$ and 0.04) compounds has been shown to cause a decrease of about 80 K at Curie temperature [205]. The Curie temperature was found as T_c = 260, 176, 180 K for $x = 0, 0.02$ and 0.04 respectively. $|\Delta S_M|$ has been reported as 4.32, 1.61, 1.24 J/kg.K for $x = 0, 0.02$ and 0.04 respectively, under 1 T magnetic field change. Similarly, in $La_{0.7}Ba_{0.2}Ca_{0.1}Mn_{1-x}Sn_xO_3$ $(x = 0$ and 0.1) compounds [206], Curie temperature decreased from 310 K for $x = 0$ to 290 K for $x = 0.1$. Under 5 T magnetic field, $x = 0$; $|\Delta S_M|$ = 6.7 J/kg.K and RCP = 248 J/kg, $x = 0.1$; $|\Delta S_M|$ 3.21 J/kg.K and RCP = 237 J/kg.K were reported. Considering the obtained $|\Delta S_M|$, RCP values and temperatures which values are observed, $La_{0.7}Ba_{0.2}Ca_{0.1}Mn_{1-x}Sn_xO_3$ compound has made one of the samples sought for cooling at room temperature.

3.2.2.11 Mn-site substitution with B, Bi, Gd, In, Ru, Sb, Si, Zn, Li

Kolat [27] studied the effect of B doping on the magnetic and magnetocaloric properties in $La_{0.67}Ca_{0.33}Mn_{1-x}B_xO_3$ $(x = 0, 0.1, 0.2$ and 0.3) compounds. Curie temperature was measured at 260 and 269 K for $x = 0$ and 0.1. Interestingly, the magnetization curves for $x = 0$ and 0.1 have only one magnetic transition been observed, while the magnetization curves for $x = 0.3$ have two magnetic transitions, one of which is $T_{c1} = 246.6$ K and the other is $T_{c2} = 210.4$ K. This is explained by the presence of two different magnetic phases, confirmed by EDX analysis, in the structure for $x = 0.3$. The magnetic entropy change has decreased from the 6.1 J/kg.K for $x = 0$ to 4.5 J/kg.K for $x = 0.3$, under 3 T magnetic field change. The decrease in saturation magnetism from 93 emu/g for $x = 0$ to 67 emu/g for $x = 0.3$ was interpreted as weakening of ferromagnetism in B-doped compounds. In the structure, the substitution of non-magnetic Bi^{3+} ions with Mn^{3+} ions causes some Mn^{3+}–O–Mn^{4+} bonds to become in the form of Bi^{3+}–O–Mn^{4+} and thus weaken ferromagnetism. Gd doping in $La_{0.7}Ca_{0.15}Sr_{0.15}Mn_{1-x}Gd_xO_3$ ($x = 0.0.02$ and 0.06) compounds has been shown to reduce Curie temperature ($T_c = 338$, 211 and 203 K for $x = 0$, 0.02 and 0.06) significantly [207]. All of the samples exhibited a second-order phase transition. $|\Delta S_m|$ and RCP values were reported as 0.925 J/kg.K RCP = 40.5 J/kg for $x = 0$, 1.2 J/kg.K and 90.7 J/kg for $x = 0.03$ and 1.004 J/kg.K and 111.14 J/kg for $x = 0.06$ under 2 T magnetic field change. Laouyenne [208] examined the magnetic and magnetocaloric properties of $La_{0.8}Na_{0.2}Mn_{1-x}Bi_xO_3$ ($0 \leq x \leq 0.06$) compounds. Curie temperature was reported as 330, 320, 310 K for $x = 0$, 0.03 and 0.06 respectively. Magnetization measurements prove that compounds exhibit a second-order phase transition. $|\Delta S_m|$ and RCP were reported as 4.73 J/kg.K and 241 J/kg for $x = 0$, 4.77 J/kg.K and 218 J/kg for $x = 0.03$ and 5.2 J/kg.K and 229 J/kg for $x = 0.06$ under 5 T magnetic field change, As can be seen from the results, doping of Bi increases the magnetocaloric properties. The increase in magnetic entropy variation is explained by the change of structural parameters which result in an increase in lattice volume and which contribute to the change of magnetic entropy. The decrease in Curie temperature with increasing Bi content was concluded as weakening of ferromagnetism in structure. In the structure, the substitution of Bi^{3+} ions with Mn^{3+} ions causes Mn^{3+} density to decrease and therefore the number of Mn^{3+}–O–Mn^{4+} bonds that interact with FM decreases. Also, when the ionic radius of the Bi^{3+} (1.03 Å) is larger compared with the ionic radius of the Mn^{3+} (0.645 Å) ion, it is evident that the substitution of the Mn^{3+} ions with the B^{3+} ions will cause a change in structural parameters such as the Mn–O bond length and the Mn–O–Mn bond angle. In this case, the decrease in Curie temperature can be attributed to the narrowing of bandwidth caused by structural deformations. The Curie temperature in Sb-doped $La_{0.67}Ba_{0.33}Mn_{1-x}Sb_xO_3$ ($x = 0.01$, 0.03 and 0.07) compounds was reported as 326, 316 and 296 K, respectively, for $x = 0.01$, 0.03 and 0.007 [209]. Magnetocaloric properties were calculated using the phenomenological model of Hamadın [57]. Under 1.5 T magnetic field change, values were reported as; $|\Delta S_M| = 1.37$ J/kg.K and RCP = 69.12 J/kg for $x = 0.01$, $|\Delta S_M| = 2.26$ J/kg.K and RCP = 87.86 J/kg for $x = 0.03$ and $|\Delta S_M| = 2.74$ J/kg.K and RCP = 122.26 J/kg for $x = 0.07$. The reduction of the Curie temperature in the Sb doped alloy is due to the decrease of the concentration of Mn^{4+} ion by increasing the concentration of Sb^{5+} ions in the structure. According to $La_{0.67}^{3+}Ba_{0.33}^{2+}Mn_{0.67+x}^{3+}Mn_{0.33-2x}^{4+}Sb_x^{5+}O_3^{2-}$ equation, the substitution of Mn atoms with Sb^{5+} ions will change the ratio of Mn^{3+}/Mn^{4+} considerably. The decrease in number of ferromagnetic M^{3+}–O–Mn^{4+} bonds is expected to affect the DE mechanism. The presence of 10 electrons in the 4d shell of Sb^{5+} ions indicates that these ions can not directly contribute to magnetic interactions. As a result, with the increase in the number of Sb^{5+} ions, ferromagnetism in the structure is weakened, and Curie temperature decreaes. Similarly, $La_{0.7}Ca_{0.3}Mn_{1-x}Zn_xO_3$ (x = 0.0, 0.06, 0.08 and 0.1) compounds may be an example for the doping of non-magnetic elements [210]. As in the scase of other non-magnetic dopings, Zn doping in this compound has caused a decrease in Curie temperature (T_c = 245, 160, 100 and 70 K respectively for x = 0, 0.06, 0.08 and 0.1). From the Arrott plots, it was determined that the samples exhibited the first-order magnetic phase transition for x < 0.06, and the second-order magnetic phase transition for $x \geq 0.06$. $|\Delta S_M|$ and RCP were reported as 10.3 J/kg.K and 294 J/kg for x = 0, 5.33 J/kg.K and 364 J/kg for x = 0.06, 3.52 J/kg.K and 404 J/kg for x = 0.08 under 5 T magnetic field change. As in similar compounds, the decrease in Curie temperature was attributed to the weakening of ferromagnetism resulting from the doping of non-magnetic ions. Depending on the change in saturation magnetization and the nature of the phase transition, $|\Delta S_M|$ decreases with increasing Zn content. Due to the increase in Zn ratio and the transformation of the nature of phase transition from first-order to second-order,the increase in RCP value has made this compound interesting. The magnetocaloric properties obtained are better than many of the previously observed alloys. With the increase in Si ratio in the $La_{2/3}Ca_{1/3}Mn_{1-x}Si_xO_3$ (x = 0.05, 0.10, 0.15 and 0.20) compounds, Curie temperature decreased, and magnetic entropy change remained at a high value almost constant [211]. In the examinations, magnetic entropy change varies in the range of 4.88-5.48 J/kg.K, under 2 T magnetic field change. On the other hand, it varies in the range of 8.78-10.32 J/kg.K, under 7 T magnetic field change. The doping of Si^{4+} ions in the structure does not directly contribute to magnetic interactions. They only weaken the long-range ferromagnetic order along the $M^{3+}-O-Mn^{4+}$ bond and consequently decrease the saturation magnetization and Curie temperature. Another example of a non-magnetic cation is the $La_{0.5}Sm_{0.1}Sr_{0.4}Mn_{1-x}In_xO_3$ ($0 \leq x \leq 0.1$) compounds [212]. In this study, it was observed that the Curie temperature decreases from 310 K for x = 0 to 251 K for x = 0.1, depending on the concentration of In. It has been determined that the compound exhibits a second-order phase transition for all In concentrations. Magnetic entropy change has been reported as $|\Delta S_m|$ = 5.88, 4.5 and 3.5 J/kg.K for x = 0, 0.05 and 0.1, under 5 T magnetic field change. Under the same magnetic field change, the relative cooling power was calculated as RCP = 181.66, 193.48 and 205.91 J/kg respectively. Since In^{3+} is a non-magnetic cation, there is no direct magnetic contribution to the system. The effect of In doping on magnetic and magnetocaloric properties is indirectly. Besides, the ionic radius of the In^{3+} (0.8 Å) ion causes the change in the structural parameters by increasing In concentration compared to the Mn^{3+} (0.65 Å) ion. One of the most prominent Mn-site doping in recent years is the Ru ion. Ru is an unique that can directly influence the magnetic and conductivity properties of manganites with local spin interactions compared to other transition metal ions. The compound of $Pr_{0.5}Ca_{0.5}Mn_{1-x}Ru_xO_3$ is one of the examples in which the effect of Ru doping on magnetic and magnetocaloric properties is studied [213]. In the study conducted by Kumar and his colleagues, [213] showed that 3% Ru doping to $Pr_{0.5}Ca_{0.5}MnO_3$ compound disrupts the antiferromagnetic CO configuration and stabilizes the ferromagnetic configuration to a steady state. Curie temperature is observed to increase from $T_c = 213$ K for $x = 0.03$ to $T_c = 239$ K for $x = 0.1$, depending on the Ru concentration. Magnetic entropy change has been observed to decrease with increasing Ru concentration (4.2, 3.8 and 3.4 J/kg.K for $x = 0.03$, 0.05 and 0.1 respectively, under 5 T magnetic field change). The relative cooling power increased from 284.9 J/kg for $x = 0.03$ to 303.6 J/kg for $x = 0.1$, due to Ru concentration. Various Mn-site manganite's magnetocaloric properties are abstracted in Table 3.

Table 3. The magnetic ordering temperature and magnetocaloric parameters of the manganite materials reported in the literature.

Material	T_c (K)	ΔH (T)	$-\Delta S_M^{max}$ (J/kg.K)	RCP(S) (J/kg)	Reference No.
A-Site doping					
(La-Li)MnO₃					
$La_{0.85}Li_{0.15}MnO_3$	235	1	1.71	67.9	284
(La-Na)MnO₃					
$La_{0.9}Na_{0.1}MnO_3$	218	1	1.53	91	217
(La-K)MnO₃					
$La_{0.85}K_{0.15}MnO_3$	238	1	2.96	7.41	274
$La_{0.9}K_{0.1}Mn_0O_3$	271	1.1	1.42	40	221
(La-Ag)MnO₃					
$La_{0.7}Ag_{0.3}MnO_3$	306	1	1.35	33	217
$La_{0.8}Ag_{0.2}MnO_3$	278	1	3.40	41	217
(La-Ca)MnO₃					
$La_{0.6}Ca_{0.4}MnO_3$	269	5	6.59	221	286
$La_{0.67}Ca_{0.33}MnO_3$	260	1.5	4.30	47	217
$La_{0.8}Ca_{0.2}MnO_3$	230	1.5	5.50	72	217
(La-Sr)MnO₃					
$La_{0.6}Sr_{0.4}MnO_3$	361	1.5	1.98	68	267
$La_{0.67}Sr_{0.33}MnO_3$	354	2	1.15	88	178

$La_{0.7}Sr_{0.3}MnO_3$	369	2	1.27	29	167
$La_{0.8}Sr_{0.2}MnO_3$	303	2	2.2	35	81
(La-Ba)MnO₃					
$La_{0.8}Ba_{0.2}MnO_3$	295	5	4.15	230	243
$La_{0.9}Ba_{0.1}MnO_3$	181	1	0.88	47.1	256
(La-Bi)MnO₃					
$La_{2.94}Bi_{0.06}MnO_3$	209	1	1.58	45.66	248
(La-Cd)MnO₃					
$La_{0.7}Cd_{0.3}MnO_3$	228	1.8	2.3	--	226
$La_{0.8}Cd_{0.2}MnO_3$	155	1.35	1.01	32	217
(La-Pb)MnO₃					
$La_{0.67}Pb_{0.33}MnO_3$	360	5	4.26	292	251
$La_{0.9}Pb_{0.1}MnO_3$	235	1.35	0.65	---	217
(La-Hf)MnO₃					
$La_{0.9}Hf_{0.1}MnO_3$	240	1	1.58	48.57	279
(La-Ca-Sr)MnO₃					
$La_{0.7}Ca_{0.1}Sr_{0.2}MnO_3$	343	1	1.43	45	238
$La_{0.7}Ca_{0.25}Sr_{0.05}MnO_3$	275	5	10.5	462	217
$La_{0.75}Ca_{0.08}Sr_{0.17}MnO_3$	336	2	2.87	97.58	193
(La-Ca-Na)MnO₃					
$La_{0.5}Ca_{0.3}Na_{0.2}MnO_3$	279	4	1.83	194.3	236
$La_{0.65}Ca_{0.20}Na_{0.20}MnO_3$	315	2	2.04	170	111
$La_{0.8}Ca_{0.1}Na_{0.1}MnO_3$	300	2	3.1	91	264
(La-Ca-K)MnO₃					
$La_{0.5}Ca_{0.45}K_{0.05}MnO_3$	295.2	2	0.92	79.08	261
(La-Ba-Na)MnO₃					
$La_{0.65}Ba_{0.3}Na_{0.05}MnO_3$	310	2	1.3	28.8(1T)	219
$La_{0.7}Ba_{0.15}Na_{0.15}MnO_3$	312	0.8	0.624	10.85	86
$La_{0.75}Ba_{0.1}Na_{0.15}MnO_3$	233	2	2.26	293.1(5T)	218
(La-K-Na)MnO₃					
$La_{0.8}K_{0.1}Na_{0.1}MnO_3$	330	5	4.39	238	225
(La-K-Ag)MnO₃					
$La_{0.8}K_{0.05}Ag_{0.15}MnO_3$	300	1	1.79	42.9	230

$La_{0.8}K_{0.1}Ag_{0.1}MnO_3$	310	5	4.92	236	225
(La-Ca-Ag)MnO₃					
$La_{0.5}Ca_{0.4}Ag_{0.1}MnO_3$	230	2	2.09	95.58	234
$La_{0.6}Ca_{0.3}Ag_{0.1}MnO_3$	267	5	8.24	264	286
(La-Ba-Ag)MnO₃					
$La_{0.65}Ba_{0.3}Ag_{0.05}MnO_3$	300	2	1.43	27.8(1T)	219
$La_{0.75}Ba_{0.1}Ag_{0.15}MnO_3$	315	2	1.72	249.1(5T)	218
(La-Ba-K)MnO₃					
$La_{0.65}Ba_{0.3}K_{0.05}MnO_3$	290	2	1.34	25.6(1T)	219
$La_{0.75}Ba_{0.1}K_{0.15}MnO_3$	288	2	2.44	247.1(5T)	218
(La-Eu-Ca)MnO₃					
$La_{0.60}Eu_{0.05}Ca_{0.35}MnO_3$	227.6	1.5	5.778	--	119
(La-Ho-Ca)MnO₃					
$(La_{0.8}Ho_{0.2})_{2/3}Ca_{1/3}MnO_3$	135	5	2.3	--	120
(La-Eu-Sr)MnO₃					
$La_{0.4}Eu_{0.3}Sr_{0.3}MnO_3$	228	5	4.55	203	288
$La_{0.6}Eu_{0.1}Sr_{0.3}MnO_3$	343	2	1.55	69	124
(La-Bi-Sr)MnO₃					
$La_{0.60}Bi_{0.05}Sr_{0.35}MnO_3$	340	1	0.98	--	239
$La_{0.7}Sr_{0.25}Bi_{0.05}MnO_3$	342	1	0.94	44	262
(La-Gd-Sr)MnO₃					
$La_{0.60}Gd_{0.05}Sr_{0.35}MnO_3$	320	1	0.81	--	239
(La-Dy-Sr)MnO₃					
$La_{0.67}Dy_{0.03}Sr_{0.3}MnO_3$	264	1	0.13	27	282
(La-Dy-Ca)MnO₃					
$La_{0.67}Dy_{0.03}Ca_{0.3}MnO_3$	240	5	3.97	157	275
$La_{0.78}Dy_{0.02}Ca_{0.2}MnO_3$	201	1	0.82	46.66	74
(La-Dy-Pb)MnO₃					
$La_{0.52}Dy_{0.15}Pb_{0.33}MnO_3$	190	5	3.51	246	251
$La_{0.63}Dy_{0.07}Pb_{0.3}MnO_3$	322	1	1.06	43	253
(La-Ce-Sr)MnO₃					
$La_{0.56}Ce_{0.14}Sr_{0.3}MnO_3$	357	2	2.6	98.8	128
(La-Nd-Sr)MnO₃					

$La_{0.57}Nd_{0.1}Sr_{0.23}MnO_3$	339	1	1.45	32.24	237
(La-Eu-Ba)MnO₃					
$La_{0.55}Eu_{0.15}Ba_{0.3}MnO_3$	298	1	2.3	46	131
$La_{0.65}Eu_{0.05}Ba_{0.3}MnO_3$	324	1	1.8	25.56	131
(La-Ca-Ba) MnO₃					
$La_{0.7}Ca_{0.18}Ba_{0.12}MnO_3$	298	1	1.85	45	217
$La_{0.7}Ca_{0.06}Ba_{0.24}MnO_3$	320	1	1.72	44	217
(La-Ca-Bi) MnO₃					
$La_{0.55}Bi_{0.15}Ca_{0.3}MnO_3$	230	7	4.41	190(3 T)	122
(La-Sm-Ca) MnO₃					
$La_{0.6}Sm_{0.01}Ca_{0.3}MnO_3$	150	0.5	0.85	--	118
(La-Sm-K) MnO₃					
$La_{0.765}Sm_{0.085}K_{0.15}MnO_3$	170	1	1.06	43.32	274
(La-Sm-Pb) MnO₃					
$La_{0.469}Sm_{0.201}Pb_{0.33}MnO_3$	286	1	1.02	--	130
$La_{0.603}Sm_{0.067}Pb_{0.33}MnO_3$	341	1	1.64	--	130
La-Sm-Sr) MnO₃					
$La_{0.5}Sm_{0.1}Sr_{0.4}MnO_3$	319	2	3.33	82.08	212
(La-Ca-Pb)MnO₃					
$La_{0.7}Ca_{0.2}Pb_{0.1}MnO_3$	327	0.01	1.022	0.322	272
(La-Pb-Na)MnO₃					
$La_{0.8}Pb_{0.1}Na_{0.1}MnO_3$	247	1	0.68	32	92
(La-Bi-Sr)MnO₃					
$La_{0.6}Bi_{0.1}Sr_{0.3}MnO_3$	334	5	4.81	226	132
(La-Y-Ca)MnO₃					
$La_{0.6}Y_{0.07}Ca_{0.33}MnO_3$	230	3	1.46	140	217
(La-Nd-Ca)MnO₃					
$La_{0.6}Nd_{0.1}Ca_{0.3}MnO_3$	233	1	1.95	37	217
(La-Pr-K)MnO₃					
$La_{0.595}Pr_{0.255}K_{0.15}MnO_3$	183	1	1.65	74.59	269
(La-Pr-K)MnO₃					
$La_{0.765}Pr_{0.085}K_{0.15}MnO_3$	225	1	3.58	83.97	269
(La-Pr-Ca)MnO₃					

$La_{0.57}Pr_{0.13}Ca_{0.3}MnO_3$	225	1	6.8	54	242
(La-Pr-Sr)MnO₃					
$La_{0.3}Pr_{0.5}Sr_{0.2}MnO_3$	188	5	3.8	255	280
$La_{0.45}Pr_{0.2}Sr_{0.35}MnO_3$	344	1	1.01	35.98	246
$La_{0.6}Pr_{0.1}Sr_{0.3}MnO_3$	329	2	1.83	95.17	252
(La-Pr-Ba)MnO₃					
$La_{0.6}Pr_{0.1}Ba_{0.3}MnO_3$	94	1	0.18	15.2	30
(La-Sm-Ba)MnO₃					
$La_{0.5}Sm_{0.2}Ba_{0.3}MnO_3$	225	5	2.48	--	276
$La_{0.6}Sm_{0.1}Ba_{0.3}MnO_3$	290	5	4.22	--	276
(La-Pr-Pb)MnO₃					
$La_{0.603}Pr_{0.067}Pb_{0.33}MnO_3$	353	1	1.04	132.47	273
(La-Dy-Ba)MnO₃					
$La_{0.665}Dy_{0.035}Ba_{0.3}MnO_3$	275	1	1.35	42.26	249
(La-Nd-Ba)MnO₃					
$La_{0.65}Nd_{0.05}Ba_{0.3}MnO_3$	325	1	1.57	24	217
$La_{0.7}Nd_{0.05}Ba_{0.25}MnO_3$	293	2	2.1	83.6	281
(La-Ca-Ce)MnO₃					
$La_{0.6}Ca_{0.37}Ce_{0.03}MnO_3$	246	1	1.96	73	277
$La_{0.6}Ca_{0.31}Ce_{0.09}MnO_3$	221	1	2.41	65	277
(La-Sr-Ba)MnO₃					
$La_{0.67}Sr_{0.11}Ba_{0.22}MnO_3$	345	5	2.258	193	250
$La_{0.7}Sr_{0.15}Ba_{0.15}MnO_3$	316	2	1.27	75.74	194
(Pr-Ca)MnO₃					
$Pr_{0.7}Ca_{0.3}MnO_3$	119	1	1	46	242
(Pr-Na)MnO₃					
$Pr_{0.8}Na_{0.2}MnO_3$	115	5	5.34	253	271
(Pr-Sr)MnO₃					
$Pr_{0.5}Sr_{0.5}MnO_3$	265	1.5	1.15	59.8	235
$Pr_{0.6}Sr_{0.4}MnO_3$	320	2.5	2.3	34.5	220
$Pr_{0.7}Sr_{0.3}MnO_3$	268.7	1	3.42	59.9	285
(Pr-Ba)MnO₃					
$Pr_{0.5}Ba_{0.5}MnO_3$	228	1	1.1	35	266

$Pr_{0.67}Ba_{0.33}MnO_3$	205	1	1.34	28	222
$Pr_{0.7}Ba_{0.3}MnO_3$	188	1	2.32	49	263
$Pr_{0.8}Ba_{0.2}MnO_3$	142.5	1	0.8	44	266
(Pr-Pb)MnO$_3$					
$Pr_{0.6}Pb_{0.4}MnO_3$	254	1.35	3.68	33	217
$Pr_{0.9}Pb_{0.1}MnO_3$	150	1.35	3.91	38	217
(Pr-Bi)MnO$_3$					
$Pr_{0.8}Bi_{0.2}MnO_3$	93	5	2.1	183	265
(Nd-Sr)MnO$_3$					
$Nd_{0.5}Sr_{0.5}MnO_3$	250	4	0.49	158	231
$Nd_{0.67}Sr_{0.33}MnO_3$	206	2	0.35	87	178
(Nd-Ba)MnO$_3$					
$Nd_{0.67}Ba_{0.33}MnO_3$	145	5	3.91	265	247
(Sm-Sr)MnO$_3$					
$Sm_{0.6}Sr_{0.4}MnO_3$	115	1	2.7	38	245
(Pr-Ca-Sr)MnO$_3$					
$Pr_{0.6}Ca_{0.1}Sr_{0.3}MnO_3$	270	5	3.64	243	204
$Pr_{0.7}Ca_{0.1}Sr_{0.2}MnO_3$	207.1	1	3.44	56.7	285
(Pr-Ca-Bi)MnO$_3$					
$Pr_{0.68}Ca_{0.22}Bi_{0.1}MnO_3$	113	1	1.094	--	35
(Pr-Sr-Ba)MnO$_3$					
$Pr_{0.5}Sr_{0.4}Ba_{0.1}MnO_3$	250	1.5	1.36	58.5	235
(Pr-Sr-Na)MnO$_3$					
$Pr_{0.5}Na_{0.05}Sr_{0.45}MnO_3$	270	2	1.6	89.9	258
(Pr-Sr-K)MnO$_3$					
$Pr_{0.5}Sr_{0.45}K_{0.05}MnO_3$	313.9	2	0.87	86.95	261
(Pr-Sr-Ag)MnO$_3$					
$Pr_{0.5}Sr_{0.3}Ag_{0.2}MnO_3$	305	2	1.34	58.2	283
$Pr_{0.6}Sr_{0.3}Ag_{0.1}MnO_3$	290	1.8	1.86	100	278
$Pr_{0.7}Ba_{0.3}MnO_3$	188	1	2.32	49	263
(Pr-Pb)MnO$_3$					
$Pr_{0.6}Pb_{0.4}MnO_3$	254	1.35	3.68	33	217
$Pr_{0.9}Pb_{0.1}MnO_3$	150	1.35	3.91	38	217

$(Pr-Bi)MnO_3$					
$Pr_{0.8}Bi_{0.2}MnO_3$	93	5	2.1	183	265
$(Nd-Sr)MnO_3$					
$Nd_{0.5}Sr_{0.5}MnO_3$	250	4	0.49	158	231
$Nd_{0.67}Sr_{0.33}MnO_3$	206	2	0.35	87	178
$(Nd-Ba)MnO_3$					
$Nd_{0.67}Ba_{0.33}MnO_3$	145	5	3.91	265	247
$(Sm-Sr)MnO_3$					
$Sm_{0.6}Sr_{0.4}MnO_3$	115	1	2.7	38	245
$(Pr-Ca-Sr)MnO_3$					
$Pr_{0.6}Ca_{0.1}Sr_{0.3}MnO_3$	270	5	3.64	243	204
$Pr_{0.7}Ca_{0.1}Sr_{0.2}MnO_3$	207.1	1	3.44	56.7	285
$(Pr-Ca-Bi)MnO_3$					
$Pr_{0.68}Ca_{0.22}Bi_{0.1}MnO_3$	113	1	1.094	--	35
$(Pr-Sr-Ba)MnO_3$					
$Pr_{0.5}Sr_{0.4}Ba_{0.1}MnO_3$	250	1.5	1.36	58.5	235
$(Pr-Sr-Na)MnO_3$					
$Pr_{0.5}Na_{0.05}Sr_{0.45}MnO_3$	270	2	1.6	89.9	258
$(Pr-Sr-K)MnO_3$					
$Pr_{0.5}Sr_{0.45}K_{0.05}MnO_3$	313.9	2	0.87	86.95	261
$(Pr-Sr-Ag)MnO_3$					
$Pr_{0.5}Sr_{0.3}Ag_{0.2}MnO_3$	305	2	1.34	58.2	283
$Pr_{0.6}Sr_{0.3}Ag_{0.1}MnO_3$	290	1.8	1.86	100	278
$(Pr-Na-K)MnO_3$					
$Pr_{0.8}Na_{0.1}K_{0.1}MnO_3$	175	5	5.68	206	271
$(Pr-Sr-Li)MnO_3$					
$Pr_{0.5}Sr_{0.3}Li_{0.2}MnO_3$	290	1.5	2.17	59.8	240
$(Pr-Bi-Sr)MnO_3$					
$Pr_{0.63}Bi_{0.07}Sr_{0.3}MnO_3$	256	2	2.28	76.47	270
$(Pr-Eu-Sr)MnO_3$					
$Pr_{0.5}Eu_{0.1}Sr_{0.4}MnO_3$	270	1	1.37	47.93	153
$(Pr-Sr-Ce)MnO_3$					
$Pr_{0.5}Sr_{0.4}Ce_{0.1}MnO_3$	310	1.5	1.93	51.1	255

(Pr-Gd-Sr)MnO₃					
$Pr_{0.5}Gd_{0.1}Sr_{0.4}MnO_3$	258	1	1.23	46.02	153
(Pr-Dy-Sr)MnO₃					
$Pr_{0.5}Dy_{0.1}Sr_{0.4}MnO_3$	248	1	1.18	48.29	153
(Pr-Sm-Sr)MnO₃					
$Pr_{0.63}Sm_{0.07}Sr_{0.3}MnO_3$	211	2	3.2	84.22	270
$Sm_{0.15}Pr_{0.4}Sr_{0.45}MnO_3$	261.1	5	3.98	240	259
$Sm_{0.45}Pr_{0.1}Sr_{0.45}MnO_3$	132	5	7.14	258.82	257
(Nd-Ca-Sr)MnO₃					
$Nd_{0.5}Sr_{0.25}Ca_{0.25}MnO_3$	175	1	0.77	140(3T)	140
(Nd-Sr-Na)MnO₃					
$Nd_{0.5}Sr_{0.3}Na_{0.2}MnO_3$	299	4	6.6	173.2	236
(Nd-Sr-K)MnO₃					
$Nd_{0.5}Sr_{0.45}K_{0.05}MnO_3$	296.4	2	0.7	74.9	261
(Nd-Pr-Sr)MnO₃					
$Nd_{0.25}Pr_{0.25}Sr_{0.5}MnO_3$	170	1.35	1.65	24	217
(Nd-La-Sr)MnO₃					
$Nd_{0.4}La_{0.3}Sr_{0.3}MnO_3$	308	1	1.2	281(5T)	260
$Nd_{0.6}La_{0.1}Sr_{0.3}MnO_3$	250	1	2.58	45	141
(Sm-Eu-Sr)MnO₃					
$Sm_{0.4}Eu_{0.2}Sr_{0.4}MnO_3$	90	1	3.8	46	245
(Pr-Er-Ca-Sr)MnO₃					
$Pr_{0.58}Er_{0.02}Ca_{0.1}Sr_{0.3}MnO_3$	260	5	3.42	246	268
Mn-Site Doping					
(La-Ca)(Ti-Mn)O₃					
$La_{0.65}Ca_{0.35}Ti_{0.1}Mn_{0.9}O_3$	103	3	1.3	182	217
(La-Ca)(Mn-M)O₃ (M=Zn, Cu)					
$La_{0.67}Ca_{0.33}Mn_{0.95}Cu_{0.15}O_3$	194	1.5	2.1	85	228
$La_{0.7}Ca_{0.3}Mn_{0.94}Zn_{0.06}O_3$	160	5	5.33	364	210
(La-Ca)(Mn-V)O₃					
$La_{0.65}Ca_{0.35}Mn_{0.9}V_{0.1}O_3$	258	5	5.25	125	202
(La-Ca)(Mn-Cr)O₃					

$La_{0.67}Ca_{0.33}Mn_{0.9}Cr_{0.1}O_3$	232.5	5	3.5	147	233
(La-Ca)(Mn-Fe)O₃					
$La_{0.8}Ca_{0.2}Mn_{0.99}Fe_{0.01}O_3$	205	5	4.32	116	180
(La-Li)(Ti-Mn)O₃					
$La_{0.85}Li_{0.15}Mn_{0.7}Ti_{0.3}O_3$	60	3	1.1	89	217
(La-Sr)(Mn-Ni)O₃					
$La_{0.7}Sr_{0.3}Mn_{0.99}Ni_{0.01}O_3$	356	1.5	2.27	--	188
(La-Ba)(Mn-Fe)O₃					
$La_{0.8}Ba_{0.2}Mn_{0.9}Fe_{0.1}O_3$	193	5	2.62	211	243
(La-Ba)(Mn-Sn)O₃					
$La_{0.67}Ba_{0.33}Mn_{0.95}Sn_{0.05}O_3$	340	2	1.9	101	203
$La_{0.67}Ba_{0.33}Mn_{0.85}Sn_{0.15}O_3$	288	2	2.49	123	203
(La-Ba)(Mn-Ti)O₃					
$La_{0.67}Ba_{0.33}Mn_{0.98}Ti_{0.02}O_3$	314	1	0.93	45	223
(La-Sr)(Mn-Ti)O₃					
$La_{0.7}Sr_{0.3}Mn_{0.9}Ti_{0.1}O_3$	210	5	2.94	288	195
$La_{0.7}Sr_{0.3}Mn_{0.92}Ti_{0.08}O_3$	235	3	2.7	135	227
(La-Sr)(Mn-Sn)O₃					
$La_{0.7}Sr_{0.3}Mn_{0.9}Sn_{0.1}O_3$	228	2	0.47	40	254
(La-Sr)(Mn-Fe)O₃					
$La_{0.67}Sr_{0.33}Mn_{0.95}Fe_{0.05}O_3$	275	5	2.8	166	164
$La_{0.7}Sr_{0.3}Mn_{0.9}Fe_{0.1}O_3$	260	2	1.7	83	224
(La-Sr)(Mn-Cu)O₃					
$La_{0.7}Sr_{0.3}Mn_{0.9}Cu_{0.1}O_3$	350	1.35	2.07	43	217
$La_{0.65}Sr_{0.35}Mn_{0.9}Cu_{0.1}O_3$	355	4	3.11	387	241
(La-Sr)(Mn-M)O₃ (M=Cr,Co)					
$La_{0.7}Sr_{0.3}Mn_{0.95}Co_{0.05}O_3$	300	1.5	1.17	46.8	244
$La_{0.7}Sr_{0.3}Mn_{0.9}Cr_{0.1}O_3$	326	2	1.76	74	254
$La_{0.7}Sr_{0.3}Mn_{0.8}Cr_{0.2}O_3$	286	2	1.203	59	167
(La-Pb)(Mn-Co)O₃					
$La_{0.67}Pb_{0.33}Mn_{0.85}Co_{0.15}O_3$	297	1	2.73	39	161
(La-Ca-Sr)(Mn-Fe)O₃					
$La_{0.75}Ca_{0.08}Sr_{0.17}Mn_{0.925}Fe_{0.075}O_3$	268	2	1.38	149	229

$La_{0.7}Ca_{0.15}Sr_{0.15}Mn_{0.9}Fe_{0.1}O_3$	244.5	5	1.86	83	192
(La-Pr-Ba)(Mn-Fe)O₃					
$La_{0.6}Pr_{0.1}Ba_{0.3}Mn_{0.9}Fe_{0.1}O_3$	277	1	0.49	25	30
(La-Pr-Sr)(Mn-Fe)O₃					
$La_{0.6}Pr_{0.1}Sr_{0.3}Mn_{0.9}Fe_{0.1}O_3$	205	2	1.15	43.12	252
(La-Ba-Sr)(Mn-Fe)O₃					
$La_{0.67}Ba_{0.22}Sr_{0.11}Mn_{0.9}Fe_{0.1}O_3$	190	5	2.261	153	250
(La-Ba-Sr)(Mn-Ga)O₃					
$La_{0.7}Ba_{0.15}Sr_{0.15}Mn_{0.9}Ga_{0.1}O_3$	301	2	1.16	72.35	194
(La-Ba-Ca)(Mn-Co)O₃					
$La_{0.8}Ba_{0.1}Ca_{0.1}Mn_{0.9}Co_{0.1}O_3$	214	5	0.8	57	165
(La-Ca-Sr)(Mn-Ga)O₃					
$La_{0.75}Ca_{0.08}Sr_{0.17}Mn_{0.9}Ga_{0.1}O_3$	241	2	1.57	101	193
(La-Nd-Sr)(Mn-Sn)O₃					
$La_{0.57}Nd_{0.1}Sr_{0.33}Mn_{0.9}Sn_{0.1}O_3$	224	5	3.22	56	204
(La-Sm-Sr)(Mn-In)O₃					
$La_{0.5}Sm_{0.1}Sr_{0.4}Mn_{0.9}In_{0.1}O_3$	261	2	2.41	120.98	212
(Pr-Ca-Sr)(Mn-Fe)O₃					
$Pr_{0.65}Ca_{0.1}Sr_{0.3}Mn_{0.95}Fe_{0.05}O_3$	185	5	3.70	233	204
(Pr-Ba-Sr)(Mn-Fe)O₃					
$Pr_{0.67}Ba_{0.22}Sr_{0.11}Mn_{0.95}Fe_{0.05}O_3$	134	5	1.90	175	287
(Pr-Ca)(Mn-Ru)O₃					
$Pr_{0.5}Ca_{0.5}Mn_{0.9}Ru_{0.1}O_3$	239	5	3.4	303(5T)	213
(Pr-Ca)(Mn-Cr)O₃					
$Pr_{0.7}Ca_{0.3}Mn_{0.8}Cr_{0.2}O_3$	155	5	2.35	253.2	232
(Pr-Ca)(Mn-Co)O₃					
$Pr_{0.7}Ca_{0.33}Mn_{0.9}Co_{0.1}O_3$	116	5	3.2	300.8	166
(Pr-Ba)(Mn-Fe)O₃					
$Pr_{0.67}Ba_{0.33}Mn_{0.95}Fe_{0.05}O_3$	128	1	0.8	49	222
(Nd-Ba)(Mn-Fe)O₃					
$Nd_{0.67}Ba_{0.33}Mn_{0.98}Fe_{0.02}O_3$	134	5	2.97	242	247
Gd					
Gd	294	5	16.8	420	217

3.3 Comparison of magnetocaloric materials

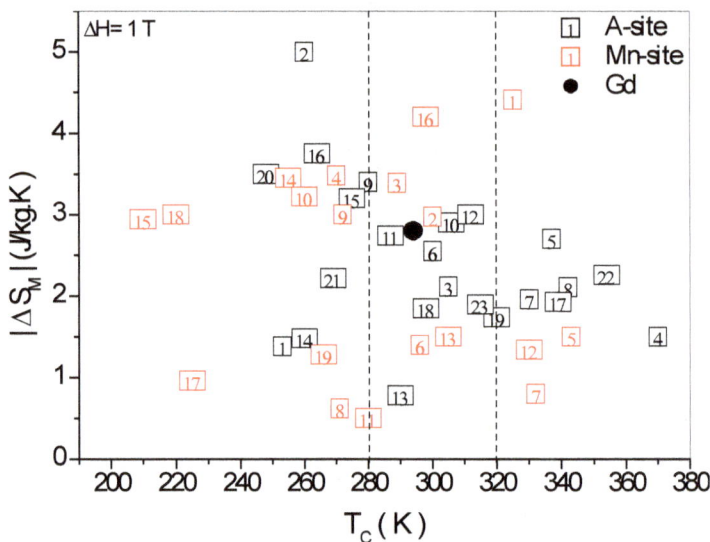

Fig.16. Comparison of Curie temperatures with maximum magnetic entropy change values of some perovskite manganite structures with Gd. (T_c = 294 K, $|\Delta S_M|$ = 2.8 J/kg.K). Black symbols, A-site doped manganites (1-$La_{0.7}Ca_{0.3}MnO_3$ 2-$La_{0.67}Ca_{0.33}MnO_{3-\delta}$ 3-$La_{0.65}Sr_{0.35}MnO_3$ 4-$La_{0.67}Sr_{0.33}MnO_3$ 5-$La_{0.67}Ba_{0.33}MnO_3$ 6-$La_{0.67}Ba_{0.33}MnO_{2.95}$ 7-$La_{0.8}Na_{0.2}MnO_3$ 8- $La_{0.835}Na_{0.165}MnO_3$ 9- $La_{0.8}Ag_{0.2}MnO_3$ 10- $La_{0.78}Ag_{0.22}MnO_3$ 11-$La_{0.9}K_{0.1}MnO_3$ 12- $La_{0.85}K_{0.15}MnO_3$ 13-$Pr_{0.55}Sr_{0.45}MnO_3$ 14- $Pr_{0.6}Sr_{0.4}MnO_3$ 15-$La_{0.67}(Ca_{0.85}Sr_{0.06})_{0.33}MnO_3$ 16- $La_{0.67}(Ca_{0.5}Sr_{0.5})_{0.33}MnO_3$ 17-$La_{0.6}Ca_{0.2}Sr_{0.2}MnO_3$ 18-$La_{0.67}Ca_{0.18}Ba_{0.12}MnO_3$ 19- $La_{0.7}Ca_{0.06}Ba_{0.24}MnO_3$ 20- $La_{0.62}Bi_{0.05}Ca_{0.33}MnO_3$ 21-$La_{0.55}Nd_{0.1}Ba_{0.35}MnO_3$ 22- $La_{0.6}Nd_{0.1}Ba_{0.3}MnO_3$ 23- $La_{0.8}Ag_{0.1}K_{0.1}MnO_3$), Red symbols, Mn-site doped manganites (1-$La_{0.77}Sr_{0.23}Mn_{0.9}Cu_{0.1}O_3$ 2-$La_{0.57}Nd_{0.1}Sr_{0.33}Mn_{0.95}Al_{0.05}O_3$ 3-$La_{0.57}Nd_{0.1}Sr_{0.33}Mn_{0.9}Al_{0.1}O_3$ 4- $La_{0.57}Nd_{0.1}Sr_{0.33}Mn_{0.85}Alu_{0.15}O_3$ 5- $La_{0.7}Sr_{0.3}Mn_{0.93}Fe_{0.07}O_3$ 6- $La_{0.7}Sr_{0.3}Mn_{0.9}Fe_{0.1}O_3$ 7- $La_{0.67}Ba_{0.33}Mn_{0.95}Fe_{0.05}O_3$ 8- $La_{0.67}Ba_{0.33}Mn_{0.9}Fe_{0.1}O_3$ 9-$La_{0.67}Pb_{0.33}Mn_{0.75}Co_{0.25}O_3$ 10- $La_{0.67}Pb_{0.33}Mn_{0.7}Co_{0.3}O_3$ 11- $Bi_{0.4}Ca_{0.6}Mn_{0.8}Ru_{0.2}O_3$ 12-$La_{0.65}Sr_{0.3}Ce_{0.05}Mn_{0.95}Cu_{0.05}O_3$ 13- $La_{0.65}Sr_{0.3}Ce_{0.05}Mn_{0.9}Cu_{0.1}O_3$ 14-$La_{0.7}Ca_{0.3}Mn_{0.9}Co_{0.1}O_3$ 15- $La_{0.7}Sr_{0.3}Mn_{0.9}Ti_{0.1}O_3$ 16- $La_{0.6}Nd_{0.1}Ca_{0.15}Sr_{0.15}Mn_{0.9}V_{0.1}O_3$ 17-$La_{0.65}Nd_{0.05}Ca_{0.3}Mn_{0.9}Cr_{0.1}O_3$ 18- $La_{0.65}Ca_{0.3}Pb_{0.05}Mn_{0.9}Cu_{0.1}O_3$ 19-$La_{0.86}Pb_{0.4}Mn_{0.9}Cu_{0.1}O_3$).

The results have revealed that the manganites could be promising with their magnetocaloric properties as magnetocaloric materials from a very low temperature to above the room temperature. In case of magnetic cooling at room temperature, the materials showing a high MC effect in the rather low field around the room temperature stand out one step further. Fig. 16 shows the magnetic entropy changes of some A-site and Mn-site manganites compared with that of the Gd element as a function of their Curie temperatures at 1 T magnetic field. It can be seen from Fig.16, MCE properties of many manganites are comparable to those of Gd. With the selection of suitable manganites and the appropriate A-site and Mn-site doping elements, the peak temperature of $|S_M|$ could be adjusted at a very wide temperature range.

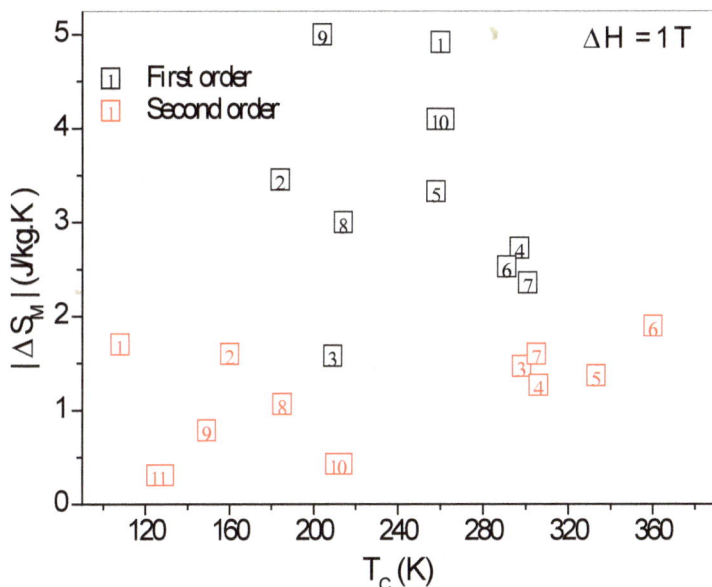

Fig. 17. Comparison of some manganite's $|\Delta S_m|$ values exhibiting the first and second-order phase transition. Black symbols; first-order phase transition (1-$La_{0.7}Ca_{0.3}MnO_3$ 2-$La_{0.6}Sm_{0.1}Ca_{0.3}MnO_3$ 3- $La_{0.94}Bi_{0.06}MnO_3$ 4- $La_{0.67}Pb_{0.33}Mn_{0.85}Co_{15}O_3$ 5- $Pr_{0.7}Sr_{0.3}MnO_3$ 6-$Pr_{0.6}Sr_{0.4}MnO_3$ 7- $Pr_{0.55}Sr_{0.45}MnO_3$ 8- $La_{0.8}Ca_{0.2}MnO_3$ 9- $La_{0.5}Pr_{0.2}Ca_{0.3}MnO_3$ 10-$La_{0.7}Ca_{0.275}Ba_{0.025}MnO_3$), red symbols; second-order phase transition (1-$La_{0.5}Sm_{0.2}Ca_{0.3}MnO_3$ 2- $La_{0.7}Ca_{0.3}Mn_{0.94}Zn_{0.06}O_3$ 3- $La_{0.6}Na_{0.1}Ca_{0.3}MnO_3$ 4-$La_{0.75}Sr_{0.1}Ca_{0.15}MnO_3$ 5- $La_{0.7}Ba_{0.3}MnO_3$ 6- $La_{0.7}Sr_{0.3}MnO_3$ 7- $La_{0.7}Ca_{0.15}Ba_{0.15}MnO_3$ 8-$La_{0.94}Bi_{0.06}Mn_{0.95}Fe_{0.05}O_3$ 9- $La_{0.8}Bi_{0.2}MnO_3$ 10- $La_{0.6}Sr_{0.4}Mn_{0.8}Fe_{0.1}Cr_{0.1}O_3$ 11-$La_{0.94}Bi_{0.06}Mn_{0.75}Cr_{0.25}O_3$).

Another important feature to be considered when selecting a magnetocaloric material is the degree of phase transition indicated by the material. The results have shown that the magnitude of the magnetocaloric effect in a material and the temperature range in which it is effective depends on the degree of phase transition of the material. Magnetic materials exhibit a first-order or second-order phase transition of around Curie temperature. Generally, materials exhibiting the first-order phase transition have a greater MC effect than materials exhibiting the second-order phase transition. In Fig.17, the maximum values of magnetic entropy changes of some manganites exhibiting the first and second-order phase the transition, and the temperatures where this value is observed are seen. It can be seen from the Fig.17, that manganites, which exhibits the first-order phase transition, has a higher $|\Delta S_M|$ value in general.

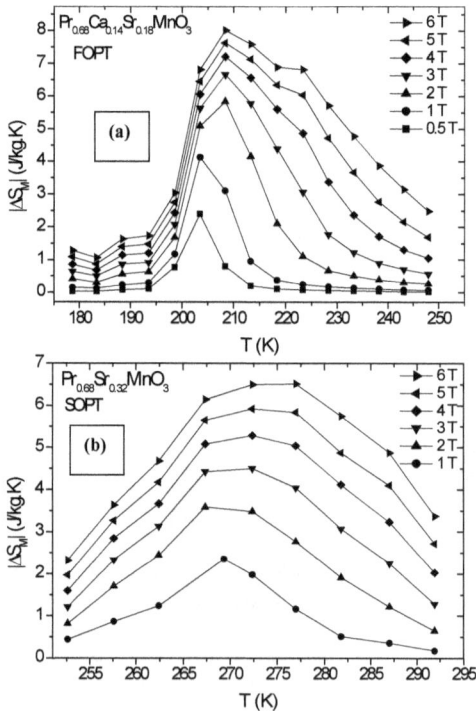

Fig. 18. Magnetic entropy changes of (a) $Pr_{0.68}Ca_{0.14}Sr_{0.18}MnO_3$ (FOPT) and (b) $Pr_{0.68}Sr_{0.32}MnO_3$ (SOPT) compounds [29].

Studies have revealed that the order of phase transition in manganites depends on the chemical composition of the compound, the type of element that is doped to the compound, ionic radius and magnetic properties. For example, $La_{0.67}Ca_{0.33}MnO_3$, exhibits the first-order phase transition but, $La_{0.67}Sr_{0.33}MnO_3$ compound exhibits the second-degree phase transition. Fig. 18 displays the magnetic entropy changes of the $Pr_{0.68}Ca_{0.14}Sr_{0.18}MnO_3$ compound, which exhibits a first-order phase transition, and the $Pr_{0.68}Sr_{0.32}MnO_3$ compound which exhibits the second-order phase transition [29]. As shown in 18.a, magnetic entropy change occurs in materials exhibiting the first-order phase transition in a relatively narrow temperature range. In materials exhibiting second-order phase transition, $|\Delta S_M|$ values are relatively smaller. However, the temperature range in which the maximum entropy change is observed is more extensive.

The RCP value calculated for $Pr_{0.68}Ca_{0.14}Sr_{0.18}MnO_3$ compound under 1 T magnetic field is 43 J/kg whereas the RCP value calculated for the $La_{0.67}Sr_{0.33}MnO_3$ compound under the same field is 142 J/kg. It can be understand from the results, relative cooling power (RCP), one of the most important parameters determining the use of magnetic materials as magnetocaloric material, is greater in the samples exhibiting the transition from the second-order phase. One of the other important points is the entropy change in the samples exhibiting second-order phase transitions is symmetrical and uniform (Fig.18). The ideal magnetic cooler for use in an Ericsson-type refrigerant must have a constant (or almost constant) magnetic entropy change in the thermodynamic cycle range. In this case, the manganites exhibiting the second-order phase transition are considered more relevant in terms of these characteristics. Also, the thermal and magnetic hysteresis losses present in materials exhibiting phase transition from the first-order are much lower than in compounds exhibiting phase transition from the second-order, making it more convenient to use these compounds as a magnetic cooler at room temperature.

Therefore, when choosing a material as a magnetic refrigerant, the nature of the phase transition exhibited by the material is one of the impotant parameters to be considered.

The literature review revealed that particle (grain) sizes in manganites have a significant effect on magnetic and magnetocaloric properties [214, 215]. A large number of studies have been carried out for this purpose. For many manganites, the effect of grain size on structural, magnetic and magnetocaloric properties has been studied in detail [70, 80, 95, 96, 120, 176, 190, 191, 214, 215]. In terms of magnetic properties, in the case of grain size is nano; spin-glass, superparamagnetic, high coercivity, saturation magnetism and Curie temperature, such as changes can be quite attractive and appealing properties. Again, it has been reported that the nature of phase transitions could be transformed from first-order to second-order with the decrease in grain size [214, 215]. As a result, since all these properties are closely related to the magnetocaloric properties of materials, it is presumed that the magnetocaloric properties will change depending on the grain size. In Fig. 19, magnetic entropy change for $La_{0.6}Ca_{0.4}MnO_3$ compound and Curie temperature change according to grain size are seen [214]. It can be seen in Fig. 19, both the $|\Delta S_M|$ value and the Curie temperature increase significantly with the extension grain size.

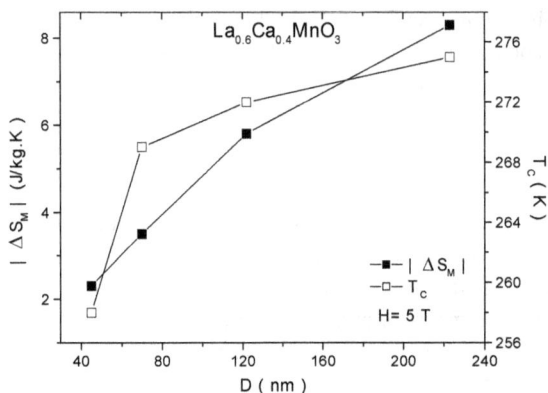

Fig.19. Variation of $|\Delta S_m|$ *and* T_c *with grain size in* $L_{0.6}Ca_{0.4}MnO_3$ *compound [214].*

The literature review showed promising candidates for magnetic cooling in an extensive temperature range of materials with magnetic entropy change in substantial quantities.

When a material is to be used as an active magnetic cooler, the first thing that comes to mind is that it has substantial $|\Delta S_M|$ values in low magnetic field changes. However, it is necessary to remember that the uniform distribution of magnetic entropy change curves plays a critical role in determining magnetic cooling efficiency [216]. Unfortunately, non-uniform magnetic entropy change (an undesirable feature for an Ericsson-circulated magnetic cooler) is a phenomenon observed in magnetocaloric materials such as Gd and many polycrystal manganites due to its non-homogenous structure [216]. In this context, in many studies [14, 103, 150, 216], it has been explained that single crystalline manganite exhibits larger magnetocaloric properties than polycrystalline manganites. Fig.20 shows the the $|\Delta S_M|$ and RCP values as a function of magnetic field changes for the single crystalline and polycrystalline forms of the $La_{0.7}Ca_{0.3}MnO_3$ compound [216]. As shown in Fig.20, the $|\Delta S_M|$ and RCP values for the single crystalline manganites are larger than the polycrystalline form.

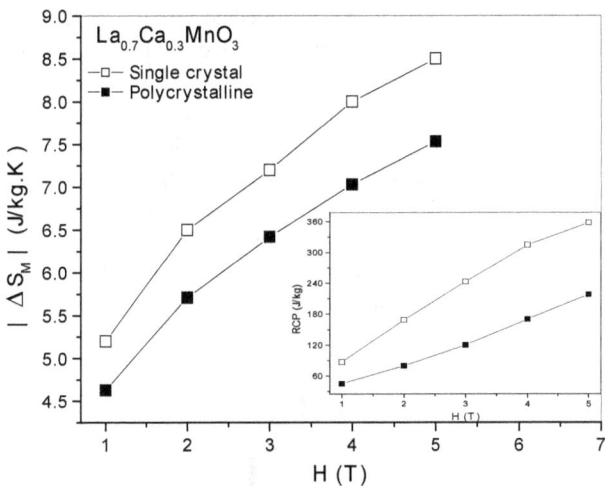

Fig.20. Variations of the $|\Delta S_m|$ and RCP values for single and polycrystalline forms of the $La_{0.7}Ca_{0.3}MnO_3$ compound, under the magnetic field [216].

Besides, asymmetric variations in magnetic entropy curves due to the effects of grain boundary in polycrystalline manganites have become more uniform and symmetrical in single crystalline manganites [14, 216]. This condition is due to the absence of grains in single crystalline manganites. Another reason for the irregularity observed in entropy curves of polycrystalline manganites is the presence of different ferromagnetic clusters originating from non-homogenous structure and stoichiometry. Most importantly, single crystalline manganites exhibit much smaller thermal and field hysteresis than polycrystalline manganites. Single crystalline manganites are more suitable magnetic refrigrant candidates than polycrystalline manganese, considering all of these results.

The literature review has revealed that different preparation methods used to produce perovskite manganites have significant effects on the structural, magnetic and magnetocaloric properties of compounds [26, 71]. To date, many methods have been applied to produce perovskite manganites. Among these, the solid-state reaction method is one of the most frequently used as a conventional method. In this method, keeping the temperature under control during a solid-state reaction is one of the most critical problem. Also, to achieve a homogeneous structure in stoichiometry, grain sizes, porosity and purity using solid-state reaction method, highly high sintering temperatures (1200-1400K) and reasonably long annealing times (> 10 h)) are required. The sol-gel method used as an alternative method is a chemical method, and it is necessary to overcome the complex chemical processes. In both methods, the production cost is quite high, and efficiency is low. In recent years, a new method called mechanical compounding or high-energy ball milling method has been introduced to produce perovskite manganites [31, 73, 74]. Studies have shown that the milling method has many advantages such as low cost, high efficiency, low-temperature synthesis and the ability to adjust grains from micrometer to nanometer degree at the desired size. In many studies, magnetic and magnetocaloric properties of manganites produced using high-energy ball milling method were investigated. In the study conducted for $La_{067}Ca_{033}MnO_3$ compound, it was observed that the perovskite structure was formed for milling time above 4 h [31].

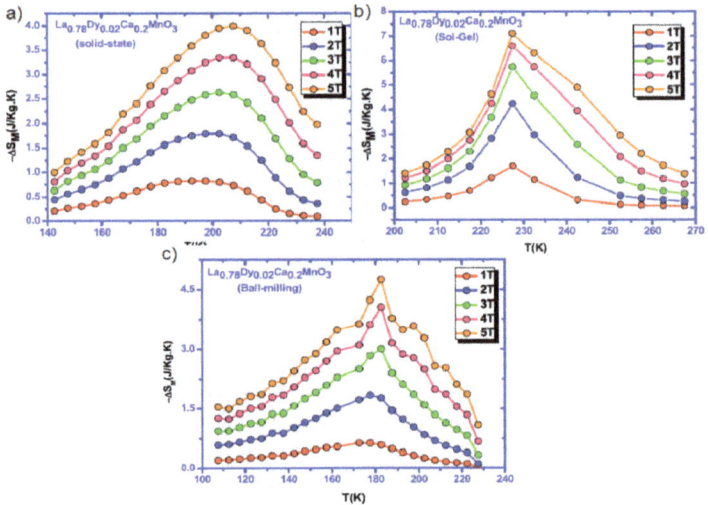

Fig. 21. Magnetic entropy changes of $La_{0.78}Dy_{0.02}Ca_{0.2}MnO_3$ compound prepared by (a) solid-state, (b) sol-gel and, (c) ball milling methods [74].

In the 24-hour milled sample, it has been reported that grain size varies from nm to a few µm. The Curie temperature, for 4, 12 and 16-hour milled samples, was reported at 250, 240 and 235 K, respectively, in $Pr_{0.5}Sr_{0.5}MnO_3$ compound, which was also produced using a ball milling method. Under 5 T magnetic field change, $|\Delta S_M|$ and RCP values were reported as 2.27, 2.57, 2.58 J/kg.K and 216.33, 214,92, 204.31 J/kg respectively for 4, 12 and 16-hour milled samples produced by milling method. Riahi [74] produced $La_{0.78}Dy_{0.02}Ca_{0.2}MnO_3$ compound using three different methods, solid-state, sol-gel and ball milling, and compared the magnetic and magnetocaloric properties of the samples. Interestingly, the samples produced by the sol-gel method exhibit the first-order phase transition, while the samples produced by the other two methods exhibit the second-order phase transition. In the sample produced by the sol-gel method, the magnetic entropy change is much more significant than the others, which is attributed to the first-order phase transition. In Fig.21, magnetic entropy changes are observed for $La_{0.78}Dy_{0.02}Ca_{0.2}MnO_3$ compound produced in three different methods.

It can be seen from Fig. 21 that the sample prepared by ball milling method has a relatively broad temperature range, although it has a smaller entropy change. Under 5 T magnetic field change, the most considerable RCP value was observed as 346.7 J/kg in the sample prepared by the ball milling method. The results proved that manganites produced by ball milling method might be promising candidates in magnetic cooling.

Conclusions

As is known, materials that exhibit a substantial temperature change under magnetic field change are called as magnetocaloric materials. A review of studies on magnetocaloric materials revealed that the temperature change of magnetocaloric materials under a magnetic field depens on many internal and external factors. Chemical composition, crystal structure and magnetic state are the most significant internal factors determining the magnetocaloric properties of a material. The most significant external factors are the temperature, the pressure of the environment and the magnitude of the field applied. To date, many materials which have larger magnetic entropy change than manganites have been reported. However, if the parameters such as preparation methods, high resistance, cost, safety in terms of health and chemical stability which determine the use of a material as an active magnetic refrigerant is considered, manganites are a good option. Considering all these features, manganites are seen to be currently the cheapest and suitable magnetic refrigerant available.

References

[1]   F. Drake, M. Purvis, and J. Hunt, Business Appreciation of Global Atmospheric Change: The United Kingdom Refrigeration Industry, Public Understanding of Science, 10, 187-211 (2001). https://doi.org/10.1088/0963-6625/10/2/303

[2]   Kyoto Protocol to the United Nations Framework Convention on Climate Change, United Nations (UN), (1997), New York, NY, USA.

[3]   Montreal Protocol on Substances that Deplete the Ozone Layer, United Nations (UN), (1987), New York, NY, USA.

[4]   A. M. Tishin, K. H. J. Buschow (1999), Handbook of Magnetic Materials, 12, 395. https://doi.org/10.1016/S1567-2719(99)12008-0

[5]   C. Zimm, A. Jastrab, A. Sternberg, V. Pecharsky, K. Gschneidner Jr.,M. Osborne, I.Anderson, Description and Performance of a Near-Room Temperature Magnetic Refrigerator, Advances in Cryogenic Engineering, 43 (1998) 1759 – 1766. https://doi.org/10.1007/978-1-4757-9047-4_222

[6]   W.F. Giauque, and D.P. MacDougall, Attainment of temperatures below 1° absolute by demagnetization of $Gd_2(SO_4)_3$ $8H_2O$, Physical Review, 43 (1933) 768. https://doi.org/10.1103/PhysRev.43.768

[7]   F.X. Hu, B.G. Shen, J.R. Sun, G.H. Wu, Large magnetic entropy change in a Heusler compound $Ni_{52.6}Mn_{23.1}Ga_{24.3}$ single crystal, Physical Review B, 64 (2001) 132412. https://doi.org/10.1103/PhysRevB.64.132412

[8]   H. Wada, Y. Tanabe, Giant magnetocaloric effect of $MnAs_{1-x}Sb_x$, Applied Physics Letters, 79 (2001) 3302. https://doi.org/10.1063/1.1419048

[9]   S. Fujieda, A. Fujita, K. Fukamichi, Large magnetocaloric effect in $La(Fe_xSi_{1-x})_{13}$ itinerant-electron metamagnetic compounds, Applied Physics Letters, 81 (2002) 1276.

[10]   V. S. Kolat, T. Izgi, H. Gencer, A. O. Kaya, N. Bayri, S. Atalay, Production of $LaFe_{11.4}Si_{1.6}$ compound at high temperature with a very short annealing time, Journal of Optoelctronics and Advanced Materials, 11 (2009),1106 – 1110. https://doi.org/10.1063/1.1498148

[11]   V. S. Kolat, A. O. Kaya, T. Izgi, H. Gencer, N. Bayri, S. Atalay, Influence of Ge and Bi substitution on the magnetic and magnetocaloric properties of LaFe11.4Si1.6, Journal of Optoelctronics and Advanced Materials,12 (2010) 1129 – 1134. Advanced Materials, 11 (2009),1106 - 1110.

[12]   T. Izgi, V. S. Kolat, H. Gencer and S. Atalay, The variation of exchange constant and magnetocaloric effect in $LaFe_{13-x}Si_x$ (x = 1.6, 1.9 and 2.2) compound International Journal of Modern Physics B, 25 (2011) 3303–3313. https://doi.org/10.1142/S0217979211101958

[13]   Q. Tegus, E. Bruck, K.H. Buschow, F.R. de Boer, Transition-metal-based magnetic refrigerants for room-temperature applications, Nature, 415(2002) 150. https://doi.org/10.1038/415150a

[14]   M.H. Phan, S.C. Yu, N.H. Hur, Excellent magnetocaloric properties of

$La_{0.7}Ca_{0.3-x}Sr_xMnO_3$ (0.05 $< x$ <0.25) single crystals, Appled Physics Letters, 86 (2005) 072504. https://doi.org/10.1063/1.1867564

[15]   Y. Sun, X. Xu, Y. Zhang, Large magnetic entropy change in the colossal magnetoresistance material $La_{2/3}Ca_{1/3}MnO_3$, Journal of Magnnetism and Magnetic Materials, 219 (2000) 183 – 185. https://doi.org/10.1016/S0304-8853(00)00433-9

[16]   H. Chen, C. Lin, D. Dai, Magnetocaloric effect in $(La,R)_{2/3}Ca_{1/3}MnO_3$ (R = Gd, Dy, Tb, Ce), Journal of Magnnetism and Magnetic Materials, 257(2003) 254 – 257. https://doi.org/10.1016/S0304-8853(02)01176-9

[17]   Y. Xu, U. Memmert ve U. Hartmann, Thermomagnetic Properties of Ferromagnetic Perovskite Manganites, Journal of Magnetism and Magnetic Materials, 242 (2002) 698 – 700. https://doi.org/10.1016/S0304-8853(01)00999-4

[18]   K.A. Gschneidner Jr, V.K. Pecharsky, The influence of magnetic field on the thermal properties of solids, Materials Science and Engginering, 287 (2000) 301–310. https://doi.org/10.1016/S0921-5093(00)00788-7

[19]   V.K. Pecharsky, K.A. Gschneidner Jr, Magnetocaloric effect from indirect measurements: magnetization and heat capacity, Journal of Applied Physics, 86 (1999) 565–575. https://doi.org/10.1063/1.370767

[20]   W. Zhong, W. Chen, W.P. Ding, N. Zhang, A. Hu, Y.W. Du, Q.J. Yan, Synthesis, structure and magnetic entropy change of polycrystalline $La_{1-x}K_xMnO_{3+\delta}$, Journal of Magnnetism and Magnetic Materials, 195 (1999) 112 – 118. https://doi.org/10.1016/S0304-8853(98)01080-4

[21]   J. Mira, J. Rivas, Drop of magnetocaloric effect related to the change from first- to second-order magnetic phase transition in $La_{2/3}(Ca_{1-x}Sr_x)_{1/3}MnO_3$, Journal of Applied Physics, 91 (2002) 8903 – 8905. https://doi.org/10.1063/1.1451892

[22]   Z.B. Guo, W. Yang, Y.T. Shen, Y.W. Du, Magnetic entropy change in $La_{0.75}Ca_{0.25-x}Sr_xMnO_3$ perovskites, Solid State Communications, 105 (1998) 89 – 92. https://doi.org/10.1016/S0038-1098(97)10064-3

[23]   H. Gencer, S. Atalay, H.I. Adiguzel, V.S. Kolat, Magnetocaloric effect in the $La_{0.62}Bi_{0.05}Ca_{0.33}MnO_3$ compound, Physica B, 357 (2005) 326–333. https://doi.org/10.1016/j.physb.2004.11.084

[24] S. Atalay, V.S. Kolat, H. Gencer, H.I. Adiguzel, Magnetic entropy-change in $La_{0.67-x}Bi_xCa_{0.33}MnO_3$ compound, Journal of Magnnetism and Magnetic Materials, 305 (2006) 452–456. https://doi.org/10.1016/j.jmmm.2006.02.082

[25] V.S. Kolat, H. Gencer, S. Atalay, Magnetic and electrical properties of $La_{0.67}Ca_{0.33}Mn_{0.9}V_{0.1}O_3$ two-phase composite, Physica B, 371 (2006) 199–204. https://doi.org/10.1016/j.physb.2005.09.036

[26] H. I. Adıgüzel, V. S. Kolat, H. Gencer, T. Seçkin, S. Atalay, Infrared, structural and magnetic properties of $La_{0.67}Ca_{0.33}MnO_3$ Compound, International Journal of Modern Physics B, 21 (2007) 43 – 53. https://doi.org/10.1142/S0217979207035935

[27] V.S. Kolat, H. Gencer, M. Gunes, S. Atalay, Effect of B-doping on the structural, magnetotransport and magnetocaloric properties of $La_{0.67}Ca_{0.33}MnO_3$ compounds, Materials Science and Engineering B, 140 (2007) 212–217. https://doi.org/10.1016/j.mseb.2007.05.002

[28] H. Gencer, M. Gunes, A. Goktas, Y. Babur, H.I. Mutlu, S. Atalay, LaBaMnO films produced by dip-coating on a quartz substrate, Journal of Alloys and Compounds, 465 (2008) 20–23. https://doi.org/10.1016/j.jallcom.2007.10.110

[29] V.S. Kolat, T. Izgi, A.O. Kaya, N. Bayri, H. Gencer, S. Atalay, Metamagnetic transition and magnetocaloric effect in charge-ordered $Pr_{0.68}Ca_{0.32-x}Sr_xMnO_3$ (x = 0,0.1,0.18,0.26, 0.32) compounds, Journal of Magnetism and Magnetic Materials, 322 (2010) 427–433. https://doi.org/10.1016/j.jmmm.2009.09.071

[30] T.Izgi, V.S.Kolat, N.Bayri, H.Gencer, S.Atalay, Structural, magnetic and magnetocaloric properties of the compound $La_{0.94}Bi_{0.06}MnO_3$, Journal of Magnetism and Magnetic Materials, 372 (2014) 112–116. https://doi.org/10.1016/j.jmmm.2014.07.037

[31] H. Gencer, N.E. Cengiz, V.S. Kolat, T. Izgi and S. Atalay, Production of $LaCaMnO_3$ composite by ball milling, Acta Physica Polonica A, 125 (2014) 214 – 216. https://doi.org/10.12693/APhysPolA.125.214

[32] M. Pektas, H. Gencer, T. Izgi, V.S. Kolat and S. Atalay, Field induced unusual magnetic behavior at low temperature in $Pr_{0.67}Ca_{0.33}MnO_3$, Acta Physica Polonica A, 125 (2014) 217 – 219. https://doi.org/10.12693/APhysPolA.125.217

[33] V.S. Kolat, U. Esturk, T. Izgi, H. Gencer , S. Atalay, The structural, magnetic and magnetocaloric properties of $La_{0.67}Ca_{0.33-x}Mg_xMnO_3$ (x = 0, 0.02, 0.05, 0.1,

0.2, 0.33) compounds, Journal of Alloys and Compounds, 628 (2015) 1–8.
https://doi.org/10.1016/j.jallcom.2014.12.129

[34]   V. S. Kolat, S. Atalay, T. Izgi, H. Gencer, N. Bayri, Structural, magnetic, and
magnetocaloric properties of $La_{1-x}Bi_xMnO_3$ (x = 0.01, 0.03, 0.06, 0.1, 0.2)
compounds, Metallurgical and Materials Transactions A, 46A (2015) 2591 –
2597. https://doi.org/10.1007/s11661-015-2839-y

[35]   H. Gencer, T. Izgi, N. Bayri, M. Pektas, V. S. Kolat, S. Atalay, Structural,
magnetic and magnetocaloric properties of $Pr_{0.68}Ca_{0.32-x}Bi_xMnO_3$ (x = 0, 0.1,
0.18, 0.26 and 0.32) compounds, Journal of Superconductivity and Novel
Magnetism, 29 (2016) 2443–2450. https://doi.org/10.1007/s10948-016-3569-0

[36]   H. Gencer, V. S. Kolat, N. Bayri, T. Izgi, and S. Atalay, Effect of Fe substitution
on magnetic and magnetocaloric properties of $La_{0.94}Bi_{0.06}Mn_{1-x}Fe_xO_3$ (x = 0, 0.05,
0.075 and 0.1) compound, Journal of Magnetics, 22 (2017) 443 – 449.
https://doi.org/10.4283/JMAG.2017.22.3.443

[37]   E.Warburg, Magnetische Untersuchungen, Annalen der Physik (Leipzig), 13
(1881) 141 – 146. https://doi.org/10.1002/andp.18812490510

[38]   E. Brück, Developments in magnetocaloric refrigeration, Journal of Physics D:
Applied Physics, 38 (2005) R381–R391. https://doi.org/10.1088/0022-
3727/38/23/R01

[39]   G.V. Brown, Magnetic heat pumping near room temperature, Journal of Applied
Physics, 47 (1976) 3673–3680. https://doi.org/10.1063/1.323176

[40]   D.T. Morelli, A.M. Mance, J.V. Mantese, A.L. Micheli, Magnetocaloric
properties of doped lanthanum manganite films, Journal of Applied Physics,79
(1996) 373 – 375. https://doi.org/10.1063/1.360840

[41]   X. X. Zhang, J. Tejada, Magnetocaloric effect in $La_{0.67}Ca_{0.33}MnO_\delta$ and
$La_{0.60}Y_{0.07}Ca_{0.33}MnO_\delta$ bulk materials, Applied Physics Letters, 69 (1996) 3596 –
3598. https://doi.org/10.1063/1.117218

[42]   V.K. Pecharsky, K.A. Gschneidner Jr, Tunable magnetic regenerator compounds
with a giant magnetocaloric effect for magnetic refrigeration from 20 to 290 K,
Appllied Physics Letters, 70 (1997) 3299 – 3301.
https://doi.org/10.1063/1.119206

[43]   A.M. Tishin, Y.I. Spichkin, The Magnetocaloric effect and its applications, Iop,
Bristol and Philadelphia 2003. https://doi.org/10.1887/0750309229

[44]   P. Weiss, R. Forrer, Magnetization of nickel and the magneto caloric effect,
        Annals of Physics (Paris), 5(1926) 153 – 213.
        https://doi.org/10.1051/anphys/192610050153

[45]   A.E. Clark, E. Callen, Cooling by adiabatic magnetization, Physical Review
        Letters, 23 (1969) 307 – 308. https://doi.org/10.1103/PhysRevLett.23.307

[46]   A.M. Tishin, Magnetocalorik effect in heavy rare earth metals and their
        compounds, PhD Thesis, Moscow State University, 1988.

[47]   S. A. Nikitin, A. S. Andreenko, A. M. Tishin, A. M. Arkharov, A. A. Zherdev,
        Magnetocaloric effect in heavy rare-earth metals, The Physics of Metal and
        Metallography, 60 (1985) 56.

[48]   B. R. Gopal, R. Chahine, T.K. Bose, A sample translatory type insert for
        automated magnetocaloric effect measurements, Review of Scientific Instrument,
        68 (1997) 1818 – 1822. https://doi.org/10.1063/1.1147999

[49]   R.D. McMichael, J.J. Ritter, R.D. Shull, Enhanced magnetocaloric effect in Gd_3
        $Ga_{5-x}Fe_xO_{12}$, Journal of Applied Physics, 73 (1993) 6946–6948.
        https://doi.org/10.1063/1.352443

[50]   K.A. Gschneidner Jr, V.K. Pecharsky, S.K. Malik, The $(Dy_{1-x}Er_x)Al_2$ compounds
        as active magnetic regenerators for magnetic refrigeration, Advances in
        Cryogenic Enginnring, 42A (1996) 475 – 482. https://doi.org/10.1007/978-1-
        4757-9059-7_63

[51]   J. O'Donnell, M. Onellion, M. S. Rzchowski, J. N. Eckstein, I. Bozovic,
        Magnetoresistance scaling in MBE-grown $La_{0.7}Ca_{0.3}MnO_3$ thin films, Physical
        Review B, 54 (1996) 6841 –4. https://doi.org/10.1103/PhysRevB.54.R6841

[52]   B. Chen, C. Uher, D.T. Orelli, J.V. Mantese, A.M. Mance, A.L. Micheli, Large
        magnetothermo power in $La_{0.67}Ca_{0.33}MnO_3$ films, Physical Review B, 53 (1995)
        5094-5097. https://doi.org/10.1103/PhysRevB.53.5094

[53]   C.M. Xiong, J.R. Sun, Y.F. Chen, B.G. Shen, J. Du, Y.X. Li, Relation between
        magnetic entropy and resistivity in $La_{0.67}Ca_{0.33}MnO_3$, IEEE Transactions on
        Magnetics, 41 (2005) 122 – 124. https://doi.org/10.1109/TMAG.2004.840132

[54]   A. Szewczyk, H. Szymczak, A. Wisniewski, K. Piotrowski, R. Kartaszynski,
        B.Dabrowski, S. Kolesnik, and Z. Bukowski, Magnetocaloric effect in
        $La_{1-x}Sr_xMnO_3$ for $x = 0.13$ and 0.16, Applied Physics Letters, 77 (2000) 1026

[55]   J.S. Amaral, N.J.O. Silva,V.S. Amaral, A mean-field scaling method for first-
        and second-order phase transition ferromagnets and its application in
        magnetocaloric studies, Appled Physics Letters, 91 (2007) 172503.
        https://doi.org/10.1063/1.2801692

[56]   J. S. Amaral, V.S. Amaral, Interpreting and modelling magnetocaloric data and
        properties: The Landau theory of phase transitions and mean-field theory, Second
        IIF-IIR International Conference on Magnetic Refrigeration at Room
        Temperature Portoroz, Slovenia, 11 – 13 April, 2007

[57]   M. A. Hamad, Magnetocaloric effect in $La_{0.65-x}Eu_xSr_{0.35}MnO_3$, Phase Transitions,
        87 (2014) 460–467. https://doi.org/10.1080/01411594.2013.828056

[58]   H. Gharsallah, M. Bejar, E. Dhahri, E.K. Hlil, L. Bessais, Prediction of
        magnetocaloric effect in $La_{0.6}Ca_{0.4-x}Sr_xMnO_3$ compounds for $x = 0$, 0.05 and 0.4
        with phenomenological model, Ceramics International, 42 (2016) 697–704.
        https://doi.org/10.1016/j.ceramint.2015.08.167

[59]   W.A. Steyert, Stirling-cycle rotating magnetic refrigerators and heat engines for
        use near room temperature, Journal of Applied Physics, 49 (1978) 1216–1226.
        https://doi.org/10.1063/1.325009

[60]   K.A. Gschneidner Jr., V.K. Pecharsky, A.O. Tsokol, Recent developments in
        magnetocaloric materials, Reports on Progress Physics, 68 (2005) 1479 – 1539.
        https://doi.org/10.1088/0034-4885/68/6/R04

[61]   G.H. Jonker, J.H. van Santen, Ferromagnetic compounds of manganese with
        perovskite structure, Physica, 16 (1950) 337–349. https://doi.org/10.1016/0031-
        8914(50)90033-4

[62]   C. Zener, Interaction between the d-shells in the transition metals, Physical
        Review, 81 (1951) 440–444. https://doi.org/10.1103/PhysRev.81.440

[63]   Z.B. Guo, Y.W. Du, J.S. Zhu, H. Huang, W.P. Ding, D. Feng, Large mag-netic
        entropy change in perovskite-type manganese oxides, Physics Review Letters, 78
        (1997) 1142 – 1145. https://doi.org/10.1103/PhysRevLett.78.1142

[64]   S.W. Cheong, Y.H. Harold, I. Y. Tokura, Colossal Magnetoresistive Oxides,
        Monographs in Condensed Matter Science, London, 1999. Gordon &Breach.

[65]   R. Kajimoto, H. Yoshizawa, Y. Tomioka, Y. Tokura, Stripe-type charge ordering
        in the metallic A-type antiferromagnet $Pr_{0.5}Sr_{0.5}MnO_3$, Physical Review B, 66
        (2002) 180402(R). https://doi.org/10.1103/PhysRevB.66.180402

[66] T. Okuda, T. Kimura, H. Kuwahara, Y. Tomioka, A. Asamitsu, Y. Okimoto, E. Saitoh, Y. Tokura, Roles of orbital in magnetoelectronic properties of colossal magnetoresistive manganites, Material Science and Engineering B, 63 (1999) 163 – 170. https://doi.org/10.1016/S0921-5107(99)00068-9

[67] M. A. Hamad, Magnetocaloric effect in $La_{1-x}Ce_xMnO_3$, Journal of Advanced Ceramics, 4 (2015) 206–210. https://doi.org/10.1007/s40145-015-0150-4

[68] A.N. Ulyanov, J.S. Kim, G.M. Shin, Y.M. Kang, S.I. Yoo, Giant magnetic entropy change in $La_{0.7}Ca_{0.3}MnO_3$ in low magnetic field, Journal of Physics D: Applied Physics, 40 (2007) 123. https://doi.org/10.1088/0022-3727/40/1/002

[69] G.C. Lin, Q. Wei, J.X. Zhang, Direct measurement of the magnetocaloric effect in $La_{0.67}Ca_{0.33}MnO_3$, Journal of Magnetism and Magnetic Materials, 300 (2006) 392 – 396. https://doi.org/10.1016/j.jmmm.2005.05.023

[70] L.E. Hueso, P. Sande, D.R. Miguens, J. Rivas, F. Rivadulla, M.A. L. Quintela, Tuning of the magnetocaloric effect in $La_{0.67}Ca_{0.33}MnO_{3-\delta}$ nanoparticles synthesized by sol–gel techniques, Journal of Applied Physics, 91 (2002) 9943 – 9347. https://doi.org/10.1063/1.1476972

[71] Z.B. Guo, J.R. Zhang, Large magnetic entropy change in $La_{0.75}Ca_{0.25}MnO_3$, Applied Physics Letter, 70 (1997) 904 – 905. https://doi.org/10.1063/1.118309

[72] A. Biswas, T. Samanta, S. Banerjee, I. Das, Inverse magnetocaloric effect in polycrystalline $La_{0.125}Ca_{0.875}MnO_3$, Journal of Physics: Condensed Matter, 21 (2009) 506005(1-3). https://doi.org/10.1088/0953-8984/21/50/506005

[73] M. Bourouina, A. Krichene, N. C. Boudjada, M. Khitouni, W. Boujelben, Structural, magnetic and magnetocaloric properties of nanostructured $Pr_{0.5}Sr_{0.5}MnO_3$ manganite synthesized by mechanical compounding , Ceramics International, 43 (2017) 8139–8145. https://doi.org/10.1016/j.ceramint.2017.03.138

[74] K. Riahi, I. Messaoui, W. C. Koubaa , S. Mercone, B. Leridon, M. Koubaa, A. Cheikhrouhou, Effect of synthesis route on the structural, magnetic and magnetocaloric properties of $La_{0.78}Dy_{0.02}Ca_{0.2}MnO_3$ manganite: A comparison between sol-gel, high-energy ball-milling and solid state process, Journal of Alloys and Compounds, 688 (2016) 1028– 1038. https://doi.org/10.1016/j.jallcom.2016.07.043

[75] M.H. Phan, S.C. Yu, N.H. Hur, Magnetic and magnetocaloric properties of $(La_{1-x})_{0.8}Ca_{0.2}MnO_3$ ($x = 0.05$, 0.20) single crystals, Journal of Magnetism and

Magnetic Materials, 262 (2003) 407 – 411. https://doi.org/10.1016/S0304-8853(03)00071-4

[76] D.L. Hou, Y. Bai, J. Xu, G.D. Tang, X.F. Nie, Magnetic entropy change in $La_{0.67-x}Ca_{0.33}MnO_3$, Journal of Alloys and Compounds, 384 (2004) 62 – 66. https://doi.org/10.1016/j.jallcom.2004.04.076

[77] W. Chen, W. Zhong, C.F. Pan, H. Chang and Y.W. Du, Curie temperature and magnetocaloric effect of polycrystalline $La_{0.8-x}Ca_{0.2}MnO_3$, Acta Physica Sinica 50 (2001) 319.

[78] A. Szewczyk, M. Gutowska, B. Dabrowski, T. Plackowski, N.P. Danilova, Y.P. Gaidukov, Specific heat anomalies in $La_{1-x}Sr_xMnO_3$ ($0.12 < x < 0.2$), Physical Review B, 71 (2005) 224432. https://doi.org/10.1103/PhysRevB.71.224432

[79] M.H. Phan, S.B. Tian, D.Q. Hoang, S.C. Yu, C. Nguyen, A.N. Ulyanov, Large magnetic-entropy change above 300 K in CMR materials, Journal of Magnetism and Magnetic Materials, 258–259 (2003) 309 – 311. https://doi.org/10.1016/S0304-8853(02)01151-4

[80] M. Pekala, V. Drozd, Magnetocaloric effect in $La_{0.8}Sr_{0.2}MnO_3$ manganite, Journal of Alloys and Compounds, 456 (2008) 30–33. https://doi.org/10.1016/j.jallcom.2007.02.092

[81] M. Pekala, V. Drozd., Magnetocaloric effect in nano- and polycrystalline $La_{0.8}Sr_{0.2}MnO_3$ manganites, Journal of Non-Crystalline Solids, 354 (2008) 5308 – 5314. https://doi.org/10.1016/j.jnoncrysol.2008.06.112

[82] M.H. Phan, S.B. Tian, S.C. Yu, A.N. Ulyanov, Magnetic and magnetocaloric properties of $La_{0.7}Ca_{0.3-x}Ba_xMnO_3$ compounds, Journal of Magnetism and Magnetic Materials, 256 (2003) 306 – 310. https://doi.org/10.1016/S0304-8853(02)00584-X

[83] W. Zhong, W. Cheng, C.T. Au, Y.W. Du, Dependence of the magnetocaloric effect on oxygen stoichiometry in polycrystalline $La_{2/3}Ba_{1/3}MnO_{3-\delta}$, Journal of Magnetism and Magnetic Materials, 261 (2003) 238 – 243. https://doi.org/10.1016/S0304-8853(02)01479-8

[84] G. Litsardakis, G.Tonozlis, The structural, magnetic and magnetocaloric properties of Ba doped La manganites, Phys. Stat. Sold. C, 11 (2014) 1133 – 1138. https://doi.org/10.1002/pssc.201300720

[85]   I.Hussain, M. S. Anwar, E. Kim, B. H. Koo, C. G. Lee, Impact of Ba Substitution on the Magnetocaloric Effect in $La_{1-x}Ba_xMnO_3$ Manganites, Korean Journal of Materials Research, 26 (2016) 623 – 627. https://doi.org/10.3740/MRSK.2016.26.11.623

[86]   S. Das, T.K. Dey, Above room temperature magnetocaloric properties of $La_{0.7}Ba_{0.3-z}Na_zMnO_3$ compounds, Materials Chemistry and Physics, 108 (2008) 220–226. https://doi.org/10.1016/j.matchemphys.2007.09.020

[87]   Y. Regaieg, M. Koubaa,W. C. Koubaa, A. Cheikhrouhou, T. Mhiri, Magnetocaloric effect above room temperature in the K-doped $La_{0.8}Na_{0.2-x}K_xMnO_3$ manganites, Journal of Alloys and Compounds, 502 (2010) 270–274. https://doi.org/10.1016/j.jallcom.2010.04.167

[88]   N.H. Luong, D.T. Hanh, N. Chau, N.D. Tho, T.D. Hiep, Properties of perovskites $La_{1-x}Cd_xMnO_3$, Journal of Magnetism and Magnetic Materials, 290–291 (2005) 690-693. https://doi.org/10.1016/j.jmmm.2004.11.338

[89]   M.A. Hamad, Magnetocaloric Effect in $La_{1-x}Cd_xMnO_3$, Journal of Superconductivity and Novel Magnetism, 26 (2013) 3459 – 3462. https://doi.org/10.1007/s10948-013-2189-1

[90]   N. Chau, H.N. Nhat, N.H. Luong, D.L. Minh, N.D. Tho, N.N. Chau, Structure, magnetic, magnetocaloric and magnetoresistance properties of $La_{1-x}Pb_xMnO_3$ perovskite, Physica B, 327 (2003) 270 – 278. https://doi.org/10.1016/S0921-4526(02)01759-3

[91]   S.G. Min, K.S. Kim, S.C. Yu, H.S. Suh, S.W. Lee, Magnetocaloric properties of $La_{1-x}Pb_xMnO_3$ ($x = 0.1, 0.2, 0.3$) compounds, IEEE Transactions on Magnics, 41 (2005) 2760. https://doi.org/10.1109/TMAG.2005.854823

[92]   A. Tozri, E. Dhahri, E.K. Hlil, (2010), Magnetic transition and magnetic entropy changes of $La_{0.8}Pb_{0.1}MnO_3$ and $La_{0.8}Pb_{0.1}Na_{0.1}MnO_3$, Materials Letters, 64 (2010) 2138 – 2141. https://doi.org/10.1016/j.matlet.2010.06.051

[93]   W. Zhong, W. Cheng, W.P. Ding, N. Zhang, Y.W. Du, Q.J. Yan, Magnetocaloric properties of Na-substituted perovskite-type manganese oxides, Solid State Communications, 106 (1998) 55 – 58. https://doi.org/10.1016/S0038-1098(97)10239-3

[94]   S. Das, T.K. Dey, Magnetocaloric effect in potassium doped lanthanum manganite perovskites prepared by a pyrophoric method, Journal of Physics: Condensed Matter, 18 (2006) 7629. https://doi.org/10.1088/0953-8984/18/32/011

[95]   S. Das, T.K. Dey, Magnetic entropy change in polycrystalline $La_{1-x}K_xMnO_3$ Perovskites, Journal of Alloys and Compounds, 440 (2007) 30–35. https://doi.org/10.1016/j.jallcom.2006.09.051

[96]   Z. Juan, L. Lirong, W. Gui, Synthesis and magnetocaloric properties of $La_{0.85}K_{0.15}MnO_3$ nanoparticles, Adv. Powder Technology, 22 (2011) 68 – 71. https://doi.org/10.1016/j.apt.2010.03.012

[97]   T. Tang, K.M. Gu, Q. Q. Cao, D.H. Wang, S.Y. Wang, S.Y. Zhang, Y.W. Du, Magnetocaloric properties of Ag-substituted perovskite-type manganites, Journal of Magnetism and Magnetic Materials, 222, (2000) 110 – 114. https://doi.org/10.1016/S0304-8853(00)00544-8

[98]   A. Coşkun, E. Taşarkuyu, A. E. Irmak, S. Aktürk, A. Ekicibil, The Structural, magnetic, and magnetocaloric properties of $La_{1-x}Ag_xMnO_3$ ($0.05 \leq x \leq 0.25$), Journal of Superconductivity and Novel Magnetics, 29 (2016) 2075 – 2084. https://doi.org/10.1007/s10948-016-3516-0

[99]   A.G. Gamzatov, A.M. Aliev, B.A. Batdalov, S.B. Abdulvagidov, O.V. Melnikov, O.Y. Gorbenko, Magnetocaloric effect in silver-doped lanthanum manganites, Technical Physics Letters, 32 (220) 471 – 473. https://doi.org/10.1134/S1063785006060046

[100]   A. M. Aliev, A. G. Gamzatov, A. B. Batdalov A. S. Mankevich, I. E. Korsakov, Structure and magnetocaloric properties of $La_{1-x}K_xMnO_3$ manganites, Physica B, 406 (2011) 885 – 889. https://doi.org/10.1016/j.physb.2010.12.021

[101]   J.M. Barandiaran, J. Gutierrez, J.R. Fernandez, M. Amboage, L. Righi, Anomalous hysteresis and metamagnetism in Bi substituted perovskites, Physica B, 343 (2004) 379–383. https://doi.org/10.1016/j.physb.2003.08.073

[102]   A. Rostamnejadi, M. Venkatesam, P. Kaneli, H. Salamati, and J.M.D. Coey, Magnetocaloric effect in $La_{0.67}Sr_{0.33}MnO_3$ manganiteabove room temperature, Journal of Magnetism and Magnetic Materials, 323 (2011) 2214–2218. https://doi.org/10.1016/j.jmmm.2011.03.036

[103]   M.H. Phan, S.C. Yu, N.H. Hur, Large magnetic entropy change above 300 K in a $La_{0.7}Ca_{0.2}Sr_{0.1}MnO_3$ single crystal, Journal of Magnetism and Magnetic Materials, 290-291 (2005) 665 – 668. https://doi.org/10.1016/j.jmmm.2004.11.330

[104] W.A. Sun, J.Q. Li, W.Q. Ao, J.N. Tang, X.Z. Gong, Hydrothermal synthesis and magnetocaloric effect of $La_{0.7}Ca_{0.2}Sr_{0.1}MnO_3$, Powder Technology, 166 (2006) 77 – 80. https://doi.org/10.1016/j.powtec.2006.05.015

[105] J.Q. Li, W.A. Sun, W.Q. Ao, J.N. Tang, Hydrothermal synthesis and Magnetocaloric effect of $La_{0.5}Ca_{0.3}Sr_{0.2}MnO_3$, Journal of Magnetism and Magnetic Materials, 302 (2006) 463 – 466. https://doi.org/10.1016/j.jmmm.2005.10.007

[106] J. S. Kim, A. N. Ulyanov, Y. M. Kang, S. G. Min, S. C. Yu, S. I. Yoo, Influence of structural-phase transition on the magnetocaloric effects of lanthanum manganites, Journal of Magnetism and Magnetic Materials, 310 (2007) 2818 – 2819. https://doi.org/10.1016/j.jmmm.2006.10.964

[107] Y. Sun, M. B. Salamon, S. H. Chun, Magnetocaloric effect and temperature coefficient of resistance of $La_{2/3}(Ca,Pb)_{1/3}MnO_3$, Journal of Applied Physics, 92 (2002) 3235. https://doi.org/10.1063/1.1502914

[108] M. H. Phan, H. X. Peng, S. C. Yu, Manganese perovskites for room temperature magnetic refrigeration applications, Journal of Magnetism and Magnetic Materials, 316 (2007) e562 – e565. https://doi.org/10.1016/j.jmmm.2007.03.021

[109] D. H. Hanh, M. S. Islam, F. A. Khan, D. L. Minh, N. Chau, Large magnetocaloric effect around room temperature in $La_{0.7}Ca_{0.3-x}Pb_xMnO_3$ perovskites, Journal of Magnetism and Magnetic Materials, 310 (2007) 2826 – 2828. https://doi.org/10.1016/j.jmmm.2006.10.1061

[110] M. Bejar, R. Dhahri, E. Dhahri, M. Balli, E.K. Hill, Large magnetic entropy change at room temperature in $La_{0.7}Ca_{0.3-x}K_xMnO_3$, Journal of Alloys and Compounds, 442 (2007) 136 – 138. https://doi.org/10.1016/j.jallcom.2006.10.170

[111] M. Koubaa, C.W. Koubaa, A. Cheikhrouhou, A.M.H. Gosnet, Structural, magnetic and magnetocaloric properties of $La_{0.65}Ca_{0.35-x}Na_xMnO_3$ Na-doped manganites, Physica B, 403 (2008) 2477 – 2483. https://doi.org/10.1016/j.physb.2008.01.011

[112] A. Mehri, C.W. Koubaa, M. Koubaa, A. Cheikh-Rouhou, (2009), Effect of sodium substitution on the structural, magnetic and magnetocaloric properties of $La_{0.5}Ca_{0.5}MnO_3$ perovskite manganites, Physics Procedia, 2 (2009) 975 – 982. https://doi.org/10.1016/j.phpro.2009.11.052

[113]   A. Mehri, W. C. R. Koubaa, M. Koubaa, A. C. Rouhou, Magnetic and
        magnetocaloric properties of monovalent substituted $La_{0.5}Ca_{0.45}A_{0.05}MnO_3$
        (A=Na, Ag, K) perovskite manganites, Materails Science and Engineering. 28
        (2012) 012049. https://doi.org/10.1088/1757-899X/28/1/012049

[114]   C.W. Koubaa, M. Koubaa, A. Cheikh-Rouhou,(2008), Structural, magneto
        transport, and magnetocaloric properties of $La_{0.7}Sr_{0.3-x}Ag_xMnO_3$ perovskite
        manganites, Journal of Alloys and Compounds, 453 (2008) 42 – 48.
        https://doi.org/10.1016/j.jallcom.2006.11.185

[115]   C.W. Koubaa, M. Koubaa, A. Cheikh-Rouhou, (2009), Effect of potassium
        doping on the structural, magnetic and magnetocaloric properties of
        $La_{0.7}Sr_{0.3-x}K_xMnO_3$ perovskite manganites, Journal of Alloys and Compounds,
        470 (2009) 42 – 46. https://doi.org/10.1016/j.jallcom.2008.03.033

[116]   G. F. Wang, Z. R. Zhao, D. L. Wang, X. F. Zhang, Tunable Curie temperature
        and magnetocaloric effect in Mg-doped (La, Sr)MnO_3 manganites, IEEE
        Transactions on Magnetics, 51 (2015) 2439732.
        https://doi.org/10.1109/INTMAG.2015.7156569

[117]   Z. M. Wang, G. Ni, Q.Y. Xu, H. Sang, Y.W. Du, Magnetocaloric effect in
        perovskite manganites $La_{0.7-x}Nd_xCa_{0.3}MnO_3$ and $La_{0.7}Ca_{0.3}MnO_3$, Journal of
        Applied Physics, 90 (2001) 5689. https://doi.org/10.1063/1.1415055

[118]   M. S. Anwar, F. Ahmed, S. R.Lee, R. Danish, B. H. Koo, Study of A-Site
        Disorder Dependent Structural, Magnetic, and Magnetocaloric Properties in
        $La_{0.7x}Sm_xCa_{0.3}MnO_3$ Manganites, Journal of Applied Physics, 52,(2013) 12.
        https://doi.org/10.7567/JJAP.52.10MC12

[119]   J. Zhang , N. Li, M. Feng , B. Pan , H. Li, Magnetic and magnetocaloric
        properties of $La_{0.65-x}Eu_xCa_{0.35}MnO_3$, Journal of Alloys and Compounds, 467
        (2009) 88–90. https://doi.org/10.1016/j.jallcom.2007.12.045

[120]   H. E. Ning, Q. I. Yang, C. Zhang, Magnetic properties and magnetocaloric
        effects of manganite $(La_{0.8}Ho_{0.2})_{2/3}Ca_{1/3}MnO_3$ and $(La_{0.5}Ho_{0.5})_{2/3}Ca_{1/3}MnO_3$,
        Journal of Rare Earth, 28 (2010) 413. https://doi.org/10.1016/S1002-
        0721(10)60282-7

[121]   M. S. Anwar, S. Kumar, F. Ahmed, N. Arshi, G. W. Kim, C. G. Lee, B. H. Koo,
        Large magnetic entropy change in $La_{0.55}Ce_{0.2}Ca_{0.25}MnO_3$ perovskite, Journal of
        Magnetics, 16 (2011) 457 – 460. https://doi.org/10.4283/JMAG.2011.16.4.457

[122]   J. Gutierrez, J. R. Fernandez, J. M. Barandiaran, I. Orue c, L. Righi, Magnetocaloric effect in $(La_{0.55}Bi_{0.15})Ca_{0.3}MnO_3$ perovskites, Sensors and Actuators A, 142 (2008) 549–553. https://doi.org/10.1016/j.sna.2007.07.025

[123]   J. S. Amaral, M. S. Reis, J. P. Araujo, P. B. Tavares, V.S. Amaral, Proceedings of the First IIF–IIR International Conference on Magnetic Refrigeration at Room Temperature, Montreux, Switzerland (2005).

[124]   A. Bouderbala, J. Makni, W. Cheikhrouhou, M. M. Koubaa, M. Koubaa, A. Cheikhtouhou, S. Nowak, S. A. Merah, Structural, magnetic and magnetocaloric study of $La_{0.7-x}Eu_xSr_{0.3}MnO_3$ (x = 0.1, 0.2 and 0.3) manganites, Ceramics International, 41 (2015) 7337 – 7344. https://doi.org/10.1016/j.ceramint.2015.02.034

[125]   V. Sudharshan, A. Saket, Magnetocaloric effect and critical field analysis in Eu substituted $La_{0.7-x}Eu_xSr_{0.3}MnO_3$ (x = 0.0, 0.1, 0.2, 0.3) manganites, Journal of Magnetism and Magnetic Materials, 446 (2018), 68 – 79. https://doi.org/10.1016/j.jmmm.2017.09.001

[126]   A.G.Gamzatov, A.M.Aliev, P.D.H. Yen, K.X. Hau, K..E. Kamaludinova, T.D. Thanh, N.T. Dung, S.C. Yu, Magnetocaloric effect in $La_{0.7-x}Pr_xSr_{0.3}MnO_3$ manganites: Direct and indirect measurements, Journal of Magnetism and Magnetic Materials, 474, (2019) 477 – 481. https://doi.org/10.1016/j.jmmm.2018.11.017

[127]   S.Phromchuai, C. Sirisathıtkul, P. Jantaratana, Effect of Gd substitution on magnetocaloric properties of lanthanum strontium manganites, Digest Journal of Nanomaterials and Biostructures, 9 (2014) 245 – 250.

[128]   S. Kallel, N. Kallel, A. Hagaza, M. Oumezzine, Large magnetic entropy change above 300 K in $(La_{0.56}Ce_{0.14})Sr_{0.3}MnO_3$ perovskite, Jounal of Alloys and Compounds, 492 (2010) 241 – 244. https://doi.org/10.1016/j.jallcom.2009.11.107

[129]   M. S. Anwar, F. Ahmed, B. H. Koo, Influence of Ce addition on the structural, magnetic, and magnetocaloric properties in $La_{0.7-x}Ce_xSr_{0.3}MnO_3$ ($0{\leq}x{\leq}0.3$) ceramic compound, Ceramics International, 41 (2015) 5821 – 5829. https://doi.org/10.1016/j.ceramint.2015.01.011

[130]   S. K. Çetin, M.Acet, M. Gunes, A. Ekicibil, M. Farle, Magnetocaloric effect in $(La_{1-x}Sm_x)_{0.67}Pb_{0.33}MnO_3$ ($0 \leq x \leq 0.3$) manganites near room temperature,

Jounal of Alloys and Compounds, 650, (2015) 285 – 294.
https://doi.org/10.1016/j.jallcom.2015.07.217

[131]   J. Dhahri, A.Dhahri, E.Dhahri, Structural, magnetic and magnetocaloric
         properties of $La_{0.7-x}Eu_xBa_{0.3}MnO_3$ perovskites, Journal of Magnetism and
         Magnetic Materials, 321 (2009) 4128–4131.
         https://doi.org/10.1016/j.jmmm.2009.08.014

[132]   S. K. Barik and R. Mahendiran, Effect of Bi doping on magnetic and
         magnetocaloric properties of $La_{0.7-x}Bi_xSr_{0.3}MnO_3$ ($0 \le x \le 0.4$), Journal of Appled
         Physics, 107 (2010) 093906(1 – 6). https://doi.org/10.1063/1.3407523

[133]   P. Chen, Y.W. Du, Large Magnetocaloric Effect in $Nd_{0.5}Sr_{0.5}MnO_3$, Chinese
         Journal of Physics, 39 (2001) 357 – 362.

[134]   L. Si, Y.L. Chang, J. Ding, C.K. Ong, B. Yao, Large magnetic entropy change in
         $Nd_{2/3}Sr_{1/3}MnO_3$, Applied Physics A, 77 (2003) 641 – 643.
         https://doi.org/10.1007/s00339-002-1537-y

[135]   R. Venkatesh, K. Sethupathi, M. Pattabiraman, G. Rangarajan. Magnetocaloric
         effect in single-crystalline $Nd_{1-x}Sr_xMnO_3$ ($x = 0.3, 0.5$), Journal of Magnetism
         and Magnetic Materials, 310 (2007) 2813 – 2814.
         https://doi.org/10.1016/j.jmmm.2006.10.1108

[136]   T. L. Phan, T.A. Ho, P.D. Thang, Q.T. Tran, T.D. Thanh, N.X. Phuc, M.H.
         Phan B.T.Huy, S.C.Yu, Critical behavior of Y-doped $Nd_{0.7}Sr_{0.3}MnO_3$
         manganites exhibiting the tricritical point and large magnetocaloric effect,
         Journal of Alloys and Compounds, 615 (2014) 937 – 945.
         https://doi.org/10.1016/j.jallcom.2014.06.107

[137]   A Beiranvand, J Tikkanen, J Rautakoski, H Huhtinen and P Paturi, Estimates of
         the magnetocaloric effect in $(Nd,Ca)MnO_3$ and $(Gd,Ca)MnO_3$ based on
         magnetic transition entropies, Materials Research Express, 4 (2017), 036101.
         https://doi.org/10.1088/2053-1591/aa5fc9

[138]   L. Dhal, E. Andharia, N. Shukla, T. G. Kumary, A.K. Nigam, S.K. Malika, P.N.
         Santhosh, R. Nirmala, Normal and inverse magnetocaloric effect in colossal
         magnetoresistive electron-doped manganites $R_{0.15}Ca_{0.85}MnO_3$ (R = Y, Gd and
         Dy), Journal of Magnetism and Magnetic Materials, 474 (2019) 215 – 220.
         https://doi.org/10.1016/j.jmmm.2018.11.003

[139] P. Sande, L.E. Hueso, D.R. Miguens, J. Rivas, F. Rivadulla, M.A. Lopez-Quintela, Large magnetocaloric effect in manganites with charge order Applied Physics Letters, 79 (2001) 2040. https://doi.org/10.1063/1.1403317

[140] J. Fan, B. Hong, D. Lu, Y. Shi, L. Ling, L. Zhang, W. Tong, L. Pi, Y. Zhang, Magnetocaloric effect of half-doped manganite $Nd_{0.5}Ca_{0.25}Sr_{0.25}MnO_3$, Physica B, 405 (2010) 3120 – 3123. https://doi.org/10.1016/j.physb.2010.04.030

[141] J. Fan, L. Ling, B. Hong, L. Pi, Y. Zhang, Magnetocaloric effect in perovskite manganite $Nd_{0.6}La_{0.1}Sr_{0.3}MnO_3$, Journal of Magnetism and Magnetic Materials, 321 (2009) 2838 – 2841. https://doi.org/10.1016/j.jmmm.2009.04.027

[142] D. Nanto, Z. Peng, Y. Y. Song, S.C. Yu, S. Telegin, L. Elochina, A. Telegin, Magnetocaloric Effect and Refrigerant Capacity of Non Stoichiometric $Nd_{0.5}Sr_{0.5}MnO_3$ Single Crystalline, IEEE Transactions on Magnetics, 48 (2012) 3995 – 3998. https://doi.org/10.1109/TMAG.2012.2208183

[143] S. Cao, J. Zhang, S. Yuan, B. Kang, C. Jing, J. Zhang, Magnetocaloric properties in the Sr-doped $Eu_{1-x}Sr_xMnO_3$ ($0.5 \leq x \leq 0.8$) system, Solid State Communications, 151 (2011) 1179–1181. https://doi.org/10.1016/j.ssc.2011.04.024

[144] A. O. Ayaş, M. Akyol, A. Ekicibil, Structural and magnetic properties with large reversible magnetocaloric effect in $(La_{1-x}Pr_x)_{0.85}Ag_{0.15}MnO_3$ ($0.0 \leq x \leq 0.5$) compounds, Philosophical Magezine, 96 (2016) 922 – 937. https://doi.org/10.1080/14786435.2016.1144939

[145] P. Chen, Y.W. Du, G. Ni, Low-field magnetocaloric effect in $Pr_{0.5}Sr_{0.5}MnO_3$, Europhysics Letters, 52 (2000) 589. https://doi.org/10.1209/epl/i2000-00478-8

[146] P. Chen, Y.W. Du, Large magnetic entropy change in $(Pr_{1-x}Nd_x)_{0.5}Sr_{0.5}MnO_3$, Journal Physics D: Applied Physics, 34 (2001) 1868. https://doi.org/10.1088/0022-3727/34/12/316

[147] I. Kammoun, W. C. Koubaa, W. Boujelben, A. Cheikhrouhou, Structural and magnetic properties of Bi doped in the A site of $(Pr_{1-x}Bi_x)_{0.6}Sr_{0.4}MnO_3$ ($0 \leq x \leq 0.4$) perovskite manganites, Journal of Materials Science, 43 (2008) 960–966. https://doi.org/10.1007/s10853-007-2194-5

[148] A.M. Gomes, F. Garcia, A.P. Guimaraes, M.S. Reis, V.S. Amaral, P.B. Tavares, Magnetocaloric effect of the (Pr,Ca)MnO₃ manganite at low temperatures, Journal of Magnetism and Magnetic Materials, 290-291 (2005) 694 – 696. https://doi.org/10.1016/j.jmmm.2004.11.339

[149] M.S. Reis, A.M. Gomes, J.P. Araujo, J.S. Amaral, P.B. Tavares, I.S. Oliveira, V.S. Amaral, Charge-ordering contribution to the magnetic entropy change of manganites, Journal of Magnetism and Magnetic Materials, 290 (2005) 697 – 699. Magnetic Materials, 290 (2005) 697 - 699.

[150] M.H. Phan, H.X. Peng, S.C. Yu, Large magnetocaloric effect in single crystal $Pr_{0.63}Sr_{0.37}MnO_3$, Journal of Applied Physics, 97 (2005) 10M306. https://doi.org/10.1063/1.1849554

[151] M.H. Phan, H.X. Peng, S.C. Yu, D.T. Hanh, N.D. Tho, N. Chau, Large magnetocaloric effect in $Pr_{1-x}Pb_xMnO_3$ ($0.1 < x < 0.5$) perovskites, Journal of Applied Physics, 99 (2006) 08Q108. https://doi.org/10.1063/1.2172212

[152] N.S. Bingham, M.H. Phan, H. Srikanth, M.A. Torija, C. Leighton, Magnetocaloric effect and refrigerant capacity in charge-ordered manganites , Journal of Applied Physics,106 (2009) 023909. https://doi.org/10.1063/1.3174396

[153] M. Moumen, A. Mehri, W. C. Koubaa, M. Koubaa, A. Cheikhrouhou, Structural, magnetic and magnetocaloric properties in $Pr_{0.5}M_{0.1}Sr_{0.4}MnO_3$ (M = Eu, Gd and Dy) polycristalline manganites, Journal of Alloys and Compounds, 509 (2011) 9084– 9088. https://doi.org/10.1016/j.jallcom.2011.06.045

[154] P. Sarkar, P. Mandal, P. Choudhury, Large magnetocaloric effect in $Sm_{0.52}Sr_{0.48}MnO_3$ in low magnetic field, Applied Physics Letters, 92 (2008) 182506. https://doi.org/10.1063/1.2919732

[155] N. S. Bingham, P. J. Lampen, T.L. Phan, M.H. Phan, S.C. Yu, H. Srikanth, Magnetocaloric effect and refrigerant capacity in $Sm_{1-x}Sr_xMnO_3$ ($x = 0.42, 0.44,$ 0.46) manganites, Journal of Applied Physics, 111 (2012) 07D705. https://doi.org/10.1063/1.3671413

[156] D. Zashchirinskii, A. Morozov, L.I. Koroleva, A. Balbashov,Magnetocaloric effect in $Sm_{0.55}Sr_{0.45}MnO_3$, Solid State Phenomenon, 168-169 (2010) 373 – 375. https://doi.org/10.4028/www.scientific.net/SSP.168-169.373

[157] Emna Tka, K. Cherif, J. Dhahri, E. Dhahri, H, Belmabrouk, E. K. Hlil, Effect of Al substitution on magnetocaloric effect in $La_{0.57}Nd_{0.1}Sr_{0.33}Mn_{1-x}Al_xO_3$ ($0.0 \leq x \leq 0.30$) polycrystalline near room temperature, Journal of Alloys and Compounds, 518 (2012) 32– 37. https://doi.org/10.1016/j.jallcom.2011.12.100

[158] P.T. Phong, L.V. Bau, L.C. Hoan, D.H. Manh, N.X. Phuc, I.J. Lee, B-site aluminum doping effect on magnetic, magnetocaloric and electro-transport properties of $La_{0.7}Sr_{0.3}Mn_{1-x}Al_xO_3$, Journal of Alloys and Compounds, 645 (2015) 243–249. https://doi.org/10.1016/j.jallcom.2015.04.225

[159] K. Dhahri, N. Dhahri, J. Dhahri, K. Taibi, E.K. Hlil, Effect of (Al, Sn) doping on structural, magnetic and magnetocaloric properties of $La_{0.7}Ca_{0.1}Pb_{0.2}Mn_{1-x-y}Al_xSn_yO_3$ ($0 < x, y < 0.075$) manganites, Journal of Alloys and Compounds, 699 (2017) 619 – 626. https://doi.org/10.1016/j.jallcom.2016.12.324

[160] Le Viet Bau, N. V. Khiem, N. X. Phuc, L. V. Hong, D. N. H. Nam, Magnetoresistance and magnetocaloric properties of $La_{0.7}Sr_{0.3}Co_{0.95}Mn_{0.05}O_3$ compound, Journal of Physics: Conference Series, 187 (2009) 012073(1 – 5). https://doi.org/10.1088/1742-6596/187/1/012073

[161] N. Dhahria, A. Dhahri, K. Cherif, J. Dhahri, H. Belmabrouk, E. Dhahri, Effect of Co substitution on magnetocaloric effect in $La_{0.67}Pb_{0.33}Mn_{1-x}Co_xO_3$ ($0.15 \le x \le 0.3$), Journal of Alloys and Compounds, 507 (2010) 405–409. https://doi.org/10.1016/j.jallcom.2010.07.222

[162] A. M. Ewasa, M. A. Hamad, Large magnetocaloric effect of $La_{0.67}Pb_{0.33}Mn_{1-x}Co_xO_3$ in small magnetic field variation, Ceramics International, 43 (2017) 7660–7662. https://doi.org/10.1016/j.ceramint.2017.03.063

[163] Y. D. Zhang, T. L. Phan, S. C. Yu, Large magnetocaloric effect for magnetic refrigeration from 210 to 275 K in $La_{0.7}Ca_{0.3}Mn_{1-x}Co_xO_3$, Journal of Applied Physics, 111 (2012) 07D703 (1-3). https://doi.org/10.1063/1.3670974

[164] P. Zhang, H.Yang,S. Zhang,H. Ge,S. Hua, Magnetic and magnetocaloric properties of perovskite $La_{0.7}Sr_{0.3}Mn_{1-x}Co_xO_3$, Physica B, 410 (2013) 1– 4. https://doi.org/10.1016/j.physb.2012.10.022

[165] M. A. Gdaiem, S. Ghodhbane, A. Dhahri, J. Dhahri, E.K. Hlil, Effect of cobalt on structural, magnetic and magnetocaloric properties of $La_{0.8}Ba_{0.1}Ca_{0.1}Mn_{1-x}Co_xO_3$ ($x = 0.00$, 0.05 and 0.10) manganites, Journal of Alloys and Compounds, 681 (2016) 547-554. https://doi.org/10.1016/j.jallcom.2016.04.143

[166] A. Selmi, R.Mnassri, W.C. Koubaa, N.C. Boudjada, A.Cheikhrouhou, The effect of Co doping on the magnetic and magnetocaloric properties of $Pr_{0.7}Ca_{0.3}Mn_{1-x}Co_xO_3$ manganites, Ceramics International, 41(2015)7723–7728. https://doi.org/10.1016/j.ceramint.2015.02.103

[167]   N. Kallel, S. Kallel, A. Hagaza, M. Oumezzine,
MagnetocaloricpropertiesintheCr-dopedLa$_{0.7}$Sr$_{0.3}$MnO$_3$ manganites, Physica B,
404 (2009) 285–288. https://doi.org/10.1016/j.physb.2008.10.049

[168]   V. S. Kumar, R. Mahendiran, A comparison of magnetocaloric effect in
Pr$_{0.6}$A$_{0.4}$Mn$_{1-x}$Cr$_x$O$_3$ (A = Ca and Sr; x = 0 and 0.04), Solid State
Communications, 150 (2010) 1445-1449.
https://doi.org/10.1016/j.ssc.2010.05.027

[169]   S. B. Abdelkhalek, N. Kallel, S. Kallel, O. Pena , M. Oumezzine, Critical
behavior and magnetic entropy change in the La$_{0.6}$Sr$_{0.4}$Mn$_{0.8}$Fe$_{0.1}$Cr$_{0.1}$O$_3$
perovskite, Journal of Magnetism and Magnetic Materials, 324 (2012) 3615–
3619. https://doi.org/10.1016/j.jmmm.2012.06.024

[170]   M. Oumezzine, O. Peña, S. Kalle, M. Oumezzine, Crossover of the
magnetocaloric effect and its importance on the determination of the critical
behaviour in the La$_{0.67}$Ba$_{0.33}$Mn$_{0.9}$Cr$_{0.1}$O$_3$ perovskite manganite, Journal of
Alloys and Compounds 539 (2012) 116–123.
https://doi.org/10.1016/j.jallcom.2012.06.043

[171]   A. Dhahri, E. Dhahri, E. K. Hlil, Structural characterization, magnetic properties
and magnetocaloric effects of La$_{0.75}$Sr$_{0.25}$Mn$_{1-x}$Cr$_x$O$_3$ (x = 0.15, 0.20, and 0.25),
Applied Physics A, 116 (2014) 2077–2085. https://doi.org/10.1007/s00339-014-
8404-5

[172]   R. Bellouz, M.Oumezzin, E.K.Hlil, E. Dhahri, Effect of Cr substitution on
magneti cand magnetic entropy change of La$_{0.65}$Eu$_{0.05}$Sr$_{0.3}$Mn$_{1-x}$Cr$_x$O$_3$ (0.05 $\leq x \leq$
0.15) rhombohedral nano rystalline near room temperature, Journal of
Magnetism and Magnetic Materials, 375(2015)136–142.
https://doi.org/10.1016/j.jmmm.2014.09.053

[173]   A. Dhahri, M. Jemmali, K. Taibi, E. Dhahri, E.K. Hlil, Structural, magnetic and
magnetocaloric properties of La$_{0.7}$Ca$_{0.2}$Sr$_{0.1}$Mn$_{1-x}$Cr$_x$O$_3$ compounds with x = 0,
0.05 and 0.1, Journal of Alloys and Compounds, 618 (2015) 488–496.
https://doi.org/10.1016/j.jallcom.2014.08.117

[174]   C. Shang, Z.C.Xia, M.Wei, Z.Jin, B.R.Chen, L.R.Shi, G.L.Xiao, S.Huang,
Tuning of Cr^{3+} ions doping on the magnetic and magnetocaloric propertiesof
La$_{0.5}$Sr$_{0.5}$Mn$_{1-x}$Cr$_x$O$_3$, Physica B, 502 (2016) 39– 47.
https://doi.org/10.1016/j.physb.2016.08.037

[175] H. Gencer, U. Ozkan, N. Bayri, T. Izgi, V. S. Kolat, Magnetic and magnetocaloric properties of $La_{0.94}Bi_{0.06}Mn_{1-x}Cr_xO_3$ (x = 0, 0.05, 0.1, 0.15, 0.2, 0.25) samples, Acta Physica Polonica 2019 (accepted for publication). https://doi.org/10.12693/APhysPolA.136.72

[176] P. Nisha, S. Savitha Pillai, A. Darbandi, M. R. Varma, K.G. Suresh, H. Hahn, Critical behaviour and magnetocaloric effect of nano crystalline $La_{0.67}Ca_{0.33}Mn_{1-x}Fe_xO_3$ (x = 0.05, 0.2) synthesized by nebulized spray pyrolysis, Materials Chemistry and Physics, 136 (2012) 66 –74. https://doi.org/10.1016/j.matchemphys.2012.06.029

[177] T. L. Phan, P. Q. Thanh, P. D. H. Yen, P. Zhang, T. D. Thanh, S.C.Yu, Ferromagnetic short-range order and magnetocaloric effect in Fe-doped $LaMnO_3$, Solid State Communications, 167 (2013) 49–53. https://doi.org/10.1016/j.ssc.2013.06.009

[178] S. Hcini, M. Boudard, S. Zemhi, M. Oumezzine, Effect of Fe-doping on structural, magnetic and magnetocaloric properties of $Nd_{0.67}Ba_{0.33}Mn_{1-x}Fe_xO_3$ manganites, Ceramics International, 40 (2014) 16041–16050. https://doi.org/10.1016/j.ceramint.2014.07.140

[179] R. B. Hassine, W. Cherif, J.A. Alonso, F. Mompean, M.T. F. Díaz , F. Elhalouani, Enhanced relative cooling power of Fe-doped $La_{0.67}Sr_{0.22}Ba_{0.11}Mn_{1-x}Fe_xO_3$ perovskites: Structural, magnetic and magnetocaloric properties, Journals of Alloys and Compounds, 649 (2015) 996 –1006. https://doi.org/10.1016/j.jallcom.2015.07.034

[180] D. Fatnassi, K. Sbissi, E. K. Hlil, M. Ellouze, J. L. Rehspringer, F. Elhalouan, Magnetic and magnetocaloric properties of nano-sized $La_{0.8}Ca_{0.2}Mn_{1-x}Fe_xO_3$ manganites prepared by sol–gel method, Journal of Nanostructur in Chemistry, 5 (2015) 375–382. https://doi.org/10.1007/s40097-015-0169-7

[181] G.F.Wang, Z.R.Zhao, H.L.Li, X.F.Zhang, Magnetocaloric effect and critical behavior in Fe-doped $La_{0.67}Sr_{0.33}Mn_{1-x}Fe_xO_3$ manganites, Ceramics International, 42 (2016) 18196–18203. https://doi.org/10.1016/j.ceramint.2016.08.138

[182] I. Betancourt, L.L. Maldonado, J.T.E. Galindo, Magnetic properties and magnetocaloric response of mixed valence $La_{2/3}Ba_{1/3}Mn_{1-x}Fe_xO_3$ manganites, Journal of Magnetism and Magnetic Materials, 401 (2016) 812–815. https://doi.org/10.1016/j.jmmm.2015.10.137

[183]   Meenakshi, A. Kumar, R. N. Mahato, Effect of Fe substitution on structural, magnetic and magnetocaloric properties of nanocrystalline $La_{0.7}Te_{0.3}Mn_{1-x}Fe_xO_3$ (x = 0.1, 0.3), Physica B, 511 (2017) 83–88. https://doi.org/10.1016/j.physb.2017.02.006

[184]   M. E. Hagary, Y.A. Shoker, M. E. Ismaila,, A.M. Moustafa, A. A. E. Aal, A.A. Ramadan, Magnetocaloric effect in manganite perovskites $La_{0.77}Sr_{0.23}Mn_{1-x}Cu_xO_3$ (0.1 ≤x ≤0.3), Solid State Communications, 149 (2009) 184 – 187. https://doi.org/10.1016/j.ssc.2008.11.023

[185]   Z. Liu, W.G. Lin, K.W. Zhou, J.L. Yan, Effect of Cu doping on the structural, magnetic and magnetocaloric properties of $La_{0.7}Sr_{0.25}Na_{0.05}Mn_{1-x}Cu_xO_3$ manganites, Ceramics International, 44 (2018) 2797–2802. https://doi.org/10.1016/j.ceramint.2017.11.021

[186]   M. Chebaane, R. Bellouz, M.Oumezzine, E. K. Hlil, A. Fouzri, Copper-doped lanthanum manganite $La_{0.65}Ce_{0.05}Sr_{0.3}Mn_{1-x}Cu_xO_3$ influence on structural, magnetic and magnetocaloric effects, RSC Advances, 8 (2018) 7186. https://doi.org/10.1039/C7RA13244A

[187]   Dwi Nanto, Seong-Cho Yu, Ralative Cooling Power og $La_{0.7}Ca_{0.3}Mn_{1-x}Cu_xO_3$ (0.0 ≤ x ≤ 0.03), International Journal of Technology, 3 (2016) 417 – 423. https://doi.org/10.14716/ijtech.v7i3.2959

[188]   Y.D. Zhang, T.L. Phan, D.S. Yang, S.C. Yu, Local structure and magnetocaloric effect for $La_{0.7}Sr_{0.3}Mn_{1-x}Ni_xO_3$, Current Applied Physics, 12 (2012) 803–807. https://doi.org/10.1016/j.cap.2011.11.010

[189]   A. Selmi, W. C. Koubaa, M. Koubaa, A. Cheikhrouhou, Effect of Ni Doping on the Structural, Magnetic and Magnetocaloric Properties of $Pr_{0.7}Ca_{0.3}Mn_{1-y}Ni_yO_3$ Manganites, Journal of Superconductivity and Novel Magnetism, 26 (2013) 1421–1428. https://doi.org/10.1007/s10948-012-1830-8

[190]   E. Oumezzine, S. Hcini, E.K. Hlil, E. Dhahri, M. Oumezzine, Effect of Ni-doping on structural, magnetic and magnetocaloric properties of $La_{0.6}Pr_{0.1}Ba_{0.3}Mn_{1-x}Ni_xO_3$ nanocrystalline manganites synthesized by Pechini sol–gel method, Journal of Alloys and Compounds, 615 (2014) 553–560. https://doi.org/10.1016/j.jallcom.2014.07.001

[191]   A. Gómez, E. Chavarriaga, J.L. Izquierdo, J. P. Gonjalc, F. Mompean, N. Rojas, O. Morán, Assessment of the relationship between magnetotransport and magnetocaloric properties in nano-sized $La_{0.7}Ca_{0.3}Mn_{1-x}Ni_xO_3$ manganites,

Journal of Magnetism and Magnetic Materials, 469 (2019) 558–569.
https://doi.org/10.1016/j.jmmm.2018.09.036

[192]   S. Othmani, M. Balli, Magnetocaloric effect and magnetic refrigeration in
$La_{0.7}Ca_{0.15}Sr_{0.15}Mn_{1-x}Ga_xO_3$ ($0 < x < 0.1$), EPJ Web of Conferences, 29 (2012)
00049. https://doi.org/10.1051/epjconf/20122900049

[193]   A. Omri, M. Bejar, M. Sajieddine, E. Dhahri, E.K. Hlil, M. E. Souni, Structural,
magnetic and magnetocaloric properties of $AMn_{1-x}Ga_xO_3$ compounds with $0 \leq x$
≤ 0.2, Physica B, 407 (2012) 2566–2572.
https://doi.org/10.1016/j.physb.2012.03.069

[194]   R. Tlili, A.Omri, M. Bekri, M. Bejar, E. Dhahri, E. K. Hlil, Effect of Ga
substitution on magnetocaloric effect in $La_{0.7}(Ba, Sr)_{0.3}Mn_{1-x}Ga_xO_3$ ($0.0 \leq x \leq$
0.20) polycrystalline at room temperature, Journal of Magnetism and Magnetic
Materials, 399 (2016) 143–148. https://doi.org/10.1016/j.jmmm.2015.09.073

[195]   S. Kallel, N. Kallel, O. Peña, M. Oumezzine, Large magnetocaloric effect in Ti-
modified $La_{0.70}Sr_{0.30}MnO_3$ perovskite, Materials Letters, 64 (2010) 1045–1048.
https://doi.org/10.1016/j.matlet.2010.02.005

[196]   P.T. Phong, L.V. Bau, L.C. Hoan, D.H. Manh, N.X. Phuc, In-Ja Lee, Effect of
B-site Ti doping on the magnetic, lowefield magnetocaloric and electrical
transport properties of $La_{0.7}Sr_{0.3}Mn_{1-x}Ti_xO_3$ perovskites, Journal of Alloys
Compounds, 656 (2016) 920 −928. https://doi.org/10.1016/j.jallcom.2015.10.038

[197]   S.E. Kossi, S.Ghodhbane, S.Mnefgui, J.Dhahri, E.K.Hlil, The impact of disorder
on magnetocaloric properties in Ti-doped manganites of $La_{0.7}Sr_{0.25}Na_{0.05}Mn_{1-x}Ti_xO_3$ ($0 \leq x \leq 0.2$), Journal of Magnetism and Magnetic Materials, 395 (2015)
134–142. https://doi.org/10.1016/j.jmmm.2015.07.050

[198]   F. B. Jemaa, S. H. Mahmood, M. Ellouz, E. K. Hlil, F. Halouani, Structural,
magnetic, magnetocaloric, and critical behavior of selected Ti-doped
manganites, Ceramics International, 41(2015) 8191–8202.
https://doi.org/10.1016/j.ceramint.2015.03.039

[199]   S. Smiy, A. Omri, R. Moussi, A. B. Ali, S. Hcini, B. F. O. Costa, E. K. Hlil, E.
Dhahri, Influence of Non-magnetic Ti^{4+} Ion Doping at Mn Site on Structural,
Magnetic, and Magnetocaloric Properties of $La_{0.5}Pr_{0.2}Sr_{0.3}Mn_{1-x}Ti_xO_3$
Manganites ($x = 0.0$ and 0.1), Journal of Superconductivity and Novel
Magnetism, (2018) 1−10. https://doi.org/10.1007/s10948-018-4825-2

[200]   A. Dhahri, F. I. H. Rhouma, S. Mnefgui, J. Dhahri, E. K. Hlil, Room temperature
critical behavior and magnetocaloric properties of
$La_{0.6}Nd_{0.1}(CaSr)_{0.3}Mn_{0.9}V_{0.1}O_3$, Ceramics International, 40 (2014) 459–464.
https://doi.org/10.1016/j.ceramint.2013.06.024

[201]   M. Mansouri, H. Omrani, R. M'nassri, W. C.Koubaa, A. Cheikhrouhou,
Vanadium-doping effects on magnetic and magnetocaloric efficiency of
$La_{0.7}Sr_{0.2}(CaLi)_{0.05}Mn_{1-x}V_xO_3$ ($x = 0.00$ and $x = 0.05$) manganites, Journal of
Materials Science: Materials in Electronics, 29 (2018) 14239–14247.
https://doi.org/10.1007/s10854-018-9557-3

[202]   A. M. Ajmi, M. Mansouri, W. C. Koubaa, M. Koubaa, A. Cheikhrouhou,
Structural, magnetic and magnetocaloric properties of vanadium-doped
manganites $La_{0.65}Ca_{0.35}Mn_{1-x}V_xO_3$ ($0 < x < 0.5$), Journal of Magnetism and
Magnetic Materials, 433 (2017) 209–215.
https://doi.org/10.1016/j.jmmm.2017.01.097

[203]   J. Dhahri, A. Dhahri, M. Oumezzine, E. Dhahri, Effect of Sn-doping on the
structural, magnetic and magnetocaloric properties of $La_{0.67}Ba_{0.33}Mn_{1-x}Sn_xO_3$
compounds, Journal of Magnetism and Magnetic Materials, 320 (2008) 2613–
2617. https://doi.org/10.1016/j.jmmm.2008.05.030

[204]   E. Tka, K. Cherif, J. Dhahri, Evolution of structural, magnetic and
magnetocaloric properties in Sn-doped manganites $La_{0.57}Nd_{0.1}Sr_{0.33}Mn_{1-x}Sn_xO_3$ ($x
= 0.05 - 0.3$), Applied Physics A, 116 (2014) 1181–1191.
https://doi.org/10.1007/s00339-013-8202-5

[205]   D. Nanto, W. Z. Nan, S.K. Oh, S.C. Yu, Influence of Sn-doping on
Magnetocaloric properties of $La_{0.7}Ca_{0.3}Mn_{1-x}Sn_xO_3$ ($x = 0.0$, $x = 0.02$ and $x =
0.04$) compound, International Journal of Technology, 3 (2016) 493 – 499.
https://doi.org/10.14716/ijtech.v7i3.2946

[206]   J. Dhahri, S. Mnefgui, A. B. Hassine, T. Tahri, M. Oumezzine, E.K. Hlil,
Behavior of the magnetocaloric effect in $La_{0.7}Ba_{0.2}Ca_{0.1}Mn_{1-x}Sn_xO_3$ manganite
oxides as promising candidates for magnetic refrigeration, Physica B, 537 (2018)
93–97. https://doi.org/10.1016/j.physb.2018.02.006

[207]   K. Riahi, A. Ezaami, I. Messaoui, M. Solzi, W. C. Koubaa, F. Cugini, G. Allodi,
F. Rossi,A. Cheikhrouhou, Investigation of the magnetic, electronic and
magnetocaloric properties of $La_{0.7}(Ca,Sr)_{0.3}Mn_{1-x}Gd_xO_3$ manganites, Journal of
Magnetism and Magnetic Materials, 441 (2017) 776–786.
https://doi.org/10.1016/j.jmmm.2017.06.091

[208] M.R. Laouyenne, M. Baazaoui, S.. Mahjoub, W. C. Koubaa,M. Oumezzine, Enhanced magnetocaloric effect with the high tunability of bismuth in $La_{0.8}Na_{0.2}Mn_{1-x}Bi_xO_3$ ($0 < x < 0.06$) perovskite manganites, Journal of Alloys and Compounds, 720 (2017) 212-220. https://doi.org/10.1016/j.jallcom.2017.05.269

[209] A. B. Hassine, A. Dhahri, L. Bouazizi, M. Oumezzine, E. K. Hlil, Characterization and theoretical investigation of the magnetocaloric effect of $La_{0.67}Ba_{0.33}Mn_{1-x}Sb_xO_3$ compounds, Solid State Communications, 233 (2016) 6–10. https://doi.org/10.1016/j.ssc.2016.02.005

[210] T.A. Ho, S.H. Lima, P.T. Thob, T.L. Phanb, S.C. Yu, Magnetic and magnetocaloric properties of $La_{0.7}Ca_{0.3}Mn_{1-x}Zn_xO_3$, Journal of Magnetism and Magnetic Materials, 426 (2017) 18–24. https://doi.org/10.1016/j.jmmm.2016.11.050

[211] L. Li, K. Nishimura, W D Hutchison, K. Mori, Large magnetocaloric effect in $La_{2/3}Ca_{1/3}Mn_{1-x}Si_xO_3$ ($x = 0.05$–0.20) manganites, Journal of Physics D: Applied Physics, 41 (2008) 175002(1−4). https://doi.org/10.1088/0022-3727/41/17/175002

[212] M. Dhahri, Asma Zaidi, K. Cherif, J. Dhahri, E. K. Hlil, Effect of indium substitution on structural, magnetic and magnetocaloric properties of $La_{0.5}Sm_{0.1}Sr_{0.4}Mn_{1-x}In_xO_3$ ($0 \leq x \leq 0.1$) manganites, Journal of Alloys and Compounds, 691 (2017) 578-586. https://doi.org/10.1016/j.jallcom.2016.08.268

[213] V. S. Kumar, R. Mahendiran, B. Raveau, Effect of Ru-Doping on Magnetocaloric Effect in Pr Based Charge Ordered Manganites, IEEE Transactions on Magnetics, 46 (2010) 1652 – 1655. https://doi.org/10.1109/TMAG.2010.2044754

[214] V.M. Andrade, R.J. Caraballo Vivas, S.S. Pedro, J.C.G. Tedesco, A.L. Rossi, A.A. Coelho, D.L. Rocco, M.S. Reis, Magnetic and magnetocaloric properties of $La_{0.6}Ca_{0.4}MnO_3$ tunable by particle size and dimensionality, Acta Materialia, 102 (2016) 49 – 55. https://doi.org/10.1016/j.actamat.2015.08.080

[215] H. Baaziz, A. Tozri, E. Dhahri, E.K. Hlil, Magnetocaloric properties of $La_{0.67}Sr_{0.33}MnO_3$ tunable by particle size and dimensionality, Chemical Physics Letters, 691 (2018) 355–359. https://doi.org/10.1016/j.cplett.2017.10.021

[216] J. C. Debnath, Novel magnetocaloric materials and room temperature magnetic refrigeration, Ph D thesis, Institute for Superconducting & Electronic Materials,

University of Wollongong, 2011. http://ro.uow.edu.au/theses/3449.
http://ro.uow.edu.au/theses/3449.

[217]   M. H. Phan, S.C. Yu, Review of the magnetocaloric effect in manganite
        materials, Journal of Magnetism and Magnetic Materials, 308 (2007) 325–340.
        https://doi.org/10.1016/j.jmmm.2006.07.025

[218] M. Koubaa, W.C.R. Koubaa, A. Cheikhrouhou, Magnetocaloric effect and
        magnetic properties of $La_{0.75}Ba_{0.1}M_{0.15}MnO_3$ (M = Na, Ag and K) perovskite
        manganites, Journal of Alloys and Compounds 479 (2009) 65–70.
        https://doi.org/10.1016/j.jallcom.2009.01.030

[219] M. Koubaa, W. C. R. Koubaa, A. Cheikhrouhou, Magnetocaloric effect in
        polycrystalline $La_{0.65}Ba_{0.3}M_{0.05}MnO_3$ (M= Na, Ag, K)manganites, Journal of
        Magnetism and Magnetic Materials, 321 (2009) 3578–3584.
        https://doi.org/10.1016/j.jmmm.2009.05.002

[220]   S. Zemni, M. Baazaoui, Ja. Dhahri, H. Vincent, M. Oumezzine, Above room
        temperature magnetocaloric effect in perovskite $Pr_{0.6}Sr_{0.4}MnO_3$, Materials
        Letters, 63 (2009) 489–491. https://doi.org/10.1016/j.matlet.2008.11.019

[221]   I. K. Kamilov, A. G. Gamzatov, A. B. Batdalov, A. S. Mankevich, and I. E.
        Korsakov, Heat Capacity and Magnetocaloric Properties of $La_{1-x}K_xMnO_3$
        Manganites, Physics of the Solid State, 52 (2010) 789-793.
        https://doi.org/10.1134/S1063783410040190

[222]   M. Baazaoui, M. Boudard, S. Zemni, Magnetocaloric properties in $Ln_{0.67}Ba_{0.33}$
        $Mn_{1-x}Fe_xO_3$ (Ln = La or Pr) manganites, Materials Letters, 65 (2011) 2093–
        2095. https://doi.org/10.1016/j.matlet.2011.04.051

[223]   M. Oumezzine, S. Zemni, O. Peña, Room temperature magnetic and
        magnetocaloric properties of $La_{0.67}Ba_{0.33}Mn_{0.98}Ti_{0.02}O_3$ perovskite, Journal of
        Alloys and Compounds (2010) 508–292.
        https://doi.org/10.1016/j.jallcom.2010.08.145

[224]   S. K. Barik, C. Krishnamoorthi, R. Mahendiran, Effect of Fe substitution on
        magnetocaloric effect in $La_{0.7}Sr_{0.3}Mn_{1-x}Fe_xO_3$ ($0.05 \leq x \leq 0.20$), Journal of
        Magnetism and Magnetic Materials, 323 (2011) 1015–1021.
        https://doi.org/10.1016/j.jmmm.2010.12.007

[225]   M. Koubaa, Y. Regaieg, W. C. Koubaa, A. Cheikhrouhou, S. A. Merah, F.
        Herbst, Magnetic and magnetocaloric properties of lanthanum manganites with

monovalent elements doping at A-site, Journal of Magnetism and Magnetic Materials, 323 (2011) 252–257. https://doi.org/10.1016/j.jmmm.2010.09.020

[226]   P. Zhang, T. L. Phan, S. C. Yu, Magnetocaloric Effect in $La_{0.7}Cd_{0.3}MnO_3$, $La_{0.7}Ba_{0.3}MnO_3$, and $Nd_{0.7}Sr_{0.3}MnO_3$, Journal of Superconductivity and Novel Magnetism, 25 (2012) 2727–2730. https://doi.org/10.1007/s10948-011-1252-z

[227]   T. L Phan, Magnetic Properties and Magnetocaloric Effect of Ti-doped $La_{0.7}Sr_{0.3}MnO_3$, Journal of the Korean Physical Society, 61 (2012) 429-433. https://doi.org/10.3938/jkps.61.429

[228]   Q. T. Phung, V. K. Vu, A. B. Ngac, H. S. Nguyen, N. N. Hoang, Magnetotransport properties and magnetocaloric effect in $La_{0.67}Ca_{0.33}Mn_{1-x}TM_xO_3$ (TM= Cu, Zn)   perovskite manganites, Journal of Magnetism and Magnetic Materials 324 (2012) 2363–2367. https://doi.org/10.1016/j.jmmm.2012.03.001

[229]   A. Omri, M. Khlifi, M. Bejar, E. Dhahri, M. Sajieddine, E.K. Hlil, Effect of Fe-doping on Magnetocaloric Properties of $AMn_{1-x}Fe_xO_3$ Compounds ($0 \leq x \leq 0.2$), Journal of Superconductivity and Novel Magnetism, 25 (2012) 1495–1500. https://doi.org/10.1007/s10948-012-1615-0

[230]   Y. Regaieg, M. Koubaa, W. Cheikhrouhou Koubaaa, A. Cheikhrouhoua, L. Sicard, S. Ammar-Merahc, F. Herbstc, Structure and magnetocaloric properties of $La_{0.8}Ag_{0.2-x}K_xMnO_3$ perovskite manganites, Materials Chemistry and Physics, 132 (2012) 839– 845. https://doi.org/10.1016/j.matchemphys.2011.12.021

[231]   A. Mehri, W. Cheikhrouhou Koubaaa, M. Koubaaa, A. Cheikhrouhou, Effect of sodium substitution on the structural, magnetic and magnetocaloric properties of $Nd_{0.5}Sr_{0.5}MnO_3$ perovskite manganites, Materials Science and Engineering, 28 (2012) 012055. https://doi.org/10.1088/1757-899X/28/1/012055

[232]   A. Selmi, W. Cheikhrouhou-Koubaa, M. Koubaaa, A. Cheikhrouhou, Structure, magnetic and magnetocloric properties of $Pr_{0.7}Ca_{0.3}Mn_{1-y}Cr_yO_3$, Materials Science and Engineering, 28 (2012) 012052. https://doi.org/10.1088/1757-899X/28/1/012052

[233]   P. Nisha, S. S. Pillai, M. R. Varma, K.G. Suresh, Critical behavior and magnetocaloric effect in $La_{0.67}Ca_{0.33}Mn_{1-x}Cr_xO_3$ ($x = 0,1, 0.25$), Solid State Sciences, 14 (2012) 40-47. https://doi.org/10.1016/j.solidstatesciences.2011.10.013

[234] M. Smari, I. Walha, E. Dhahri, E.K. Hlil, Structural, magnetic and magnetocaloric properties of Ag-doped $La_{0.5}Ca_{0.5-x}Ag_xMnO_3$ compounds with $0 \leq x \leq 0.4$, Journal of Alloys and Compounds, 579 (2013) 564–571. https://doi.org/10.1016/j.jallcom.2013.07.104

[235] S. Hua, P. Zhang, H. Yang, S. Zhang, and H. Ge, Magnetic and Magnetocaloric Properties of Perovskite $Pr_{0.5}Sr_{0.5-x}Ba_xMnO_3$, Journal of Magnetics, 18 (2013) 386-390. https://doi.org/10.4283/JMAG.2013.18.4.386

[236] A. Mehri, W. C. Koubaa, M. Koubaa and A. Cheikhrouhou, Magnetocaloric properties in $La_{0.5}Ca_{0.3}Na_{0.2}MnO_3$, $Pr_{0.5}Sr_{0.3}Na_{0.2}MnO_3$ and $Nd_{0.5}Sr_{0.3}Na_{0.2}$ manganites, Journal of the Korean Physical Society, 63 (2013) 722-725. https://doi.org/10.3938/jkps.63.722

[237] S. Mnefgui, A. Dhahri, N. Dhahri, El. K. Hlil, J. Dhahri, The effect deficient of strontiuon structural, magnetic and magnetocaloric properties of $La_{0.57}Nd_{0.1}Sr_{0.33-x}MnO_3$ ($x = 0.1$ and 0.15) manganite, Journal of Magnetism and Magnetic Materials, 340 (2013) 91–96. https://doi.org/10.1016/j.jmmm.2013.03.030

[238] M. S. Anwar, F. Ahmed, G. W. Kim, S. N. Heo and B. H. Koo, The Interplay of Ca and Sr in the Bulk Magnetocaloric $La_{0.7}Sr_{(0.3-x)}Ca_x MnO_3$ ($x = 0, 0.1$ and 0.3) Manganite, Journal of the Korean Physical Society, 62 (2013) 1974-1978. https://doi.org/10.3938/jkps.62.1974

[239] P. Zhang, H. Yang, S. Zhang, H. Ge, M. Pan, Effect of Li doping on the magnetic and magnetocaloric properties of $Pr_{0.5}Sr_{0.5-x}Li_xMnO_3$ ($0 \leq x \leq 0.3$), Journal of Magnetism and Magnetic Materials, 334 (2013) 16–20. https://doi.org/10.1016/j.jmmm.2013.01.008

[240] M. S. Anwar, S. Kumar, F. Ahmed, S. N. Heo, G. W. Kim, B. H. Koo, Study of magnetic entropy change in $La_{0.65}Sr_{0.35}Cu_{0.1}Mn_{0.9}O_3$ complex perovskite, Journal of Electroceramics, 30 (2013) 46–50. https://doi.org/10.1007/s10832-012-9711-x

[241] M. Suemitsu, T. Nakagawa, Y. Hirayama, S. Seino, T. A. Yamamoto, Magnetocaloric effect of $La_{0.7-x}Pr_xCa_{0.3}MnO_3$ perovskites, Journal of Alloys and Compounds, 551 (2013) 195–199. https://doi.org/10.1016/j.jallcom.2012.10.063

[242] S. Ghodhbane, A. Dhahri, N. Dhahri, E.K. Hlil, J. Dhahri, Structural, magnetic and magnetocaloric properties of $La_{0.8}Ba_{0.2}Mn_{1-x}Fe_xO_3$ compounds with $0 \leq x \leq 0.1$, Journal of Alloys and Compounds, 550 (2013) 358–364. https://doi.org/10.1016/j.jallcom.2012.10.087

[243]   C. P. Reshmi, S. S. Pillai, K.G. Suresh, M. R. Varma, Near room temperature
magnetocaloric properties of Fe substituted $La_{0.67}Sr_{0.33}MnO_3$, Materials Research
Bulletin, 48 (2013) 889–894. https://doi.org/10.1016/j.materresbull.2012.11.084

[244]   M. Aparnadevi, R. Mahendiran, Effect of Eu Doping on Magnetocaloric Effect
in $Sm_{0.6}Sr_{0.4}MnO_3$, Integrated Ferroelectrics, 142 (2013) 1–6.
https://doi.org/10.1080/10584587.2013.780144

[245]   M.S. Anwar, F Ahmed, B. H. Koo, Structural distortion effect on the
magnetization and magnetocaloric effect in Pr modified $La_{0.65}Sr_{0.35}MnO_3$
manganite, Journal of Alloys and Compounds, 617 (2014) 893–898.
https://doi.org/10.1016/j.jallcom.2014.08.105

[246]   G. F. Wang, L. R. Li, Z. R. Zhao, X. Q. Yu, X. F. Zhang, Structural and
magnetocaloric effect of $Ln_{0.67}Sr_{0.33}MnO_3$ (Ln = La, Pr and Nd) nanoparticles,
Ceramics International, 40 (2014) 16449–16454.
https://doi.org/10.1016/j.ceramint.2014.07.154

[247]   R. Cherif, E. K. Hlil, M. Ellouze, F. Elhalouani, S. Obbade, Study of magnetic
and magnetocaloric properties of $La_{0.6}Pr_{0.1}Ba_{0.3}MnO_3$ and
$La_{0.6}Pr_{0.1}Ba_{0.3}Mn_{0.9}Fe_{0.1}O_3$ perovskite-type manganese oxides, Journal of
Materials Science, 49 (2014) 8244 –8251. https://doi.org/10.1007/s10853-014-
8533-4

[248]   G. Tonozlis and G. Litsardakis, Magnetic and magnetocaloric properties of Dy-
substituted $(La_{1-x}Dy_x)_{0.7}Ba_{0.3}MnO_3$ Perovskites, IEEE Transactions on
Magnetics, 50 (2014) 2506404. https://doi.org/10.1109/TMAG.2014.2325745

[249]   F. B. Jemaa, S. Mahmood, M. Ellouze, E. K. Hlil, F. Halouani, I. Bsoul, M.
Awawdeh, Structural, magnetic and magnetocaloric properties of $La_{0.67}Ba_{0.22}$
$Sr_{0.11}Mn_{1-x}Fe_xO_3$ nanopowders, Solid State Sciences, 37 (2014) 121–130.
https://doi.org/10.1016/j.solidstatesciences.2014.09.004

[250]   N. Zaidi, S. Mnefgui, A. Dhahri, J. Dhahri, E. K. Hlil, The effect of Dy doped on
structural, magnetic and magnetocaloric properties of $La_{0.67-x}Dy_xPb_{0.33}MnO_3$ (x =
0.00, 0.10 and 0.20) compounds, Physica B, 450 (2014) 155–161.
https://doi.org/10.1016/j.physb.2014.05.068

[251]   S. Mahjoub, M. Baazaoui, R. M'nassri, H. Rahmouni, N. C. Boudjada, M.
Oumezzine, Effect of iron substitution on the structural, magnetic and
magnetocaloric properties of $Pr_{0.6}Ca_{0.1}Sr_{0.3}Mn_{1-x}Fe_xO_3$ ($0 \le x \le 0.075$)

manganites, Journal of Alloys and Compounds, 608 (2014) 191–196.
https://doi.org/10.1016/j.jallcom.2014.04.125

[252]   R. Cherif, E. K. Hlil, M. Ellouze, F. Elhalouani, S. Obbade, Magnetic and
magnetocaloric properties of $La_{0.6}Pr_{0.1}Sr_{0.3}Mn_{1-x}Fe_xO_3$ ($0 \leq x \leq 0.3$) manganites,
Journal of Solid State Chemistry, 215 (2014) 271–276.
https://doi.org/10.1016/j.jssc.2014.04.004

[253]   T. A. Ho, T. D. Thanh, P. D. Thang, J. S. Lee, T. L. Phan, and S. C. Yu,
Magnetic properties and magnetocaloric effect in Pb-doped $La_{0.9}Dy_{0.1}MnO_3$
manganites, IEEE Transactions on Magnetics, 50 (2014) 2502504.

[254]   B. Arayedh, S. Kallel, N. Kallel, O. Peña, Influence of non-magnetic and
magnetic ions on the Magnetocaloric properties of $La_{0.7}Sr_{0.3}Mn_{0.9}M_{0.1}O_3$ doped
in the Mn sites by M = Cr, Sn, Ti, Journal of Magnetism and Magnetic
Materials, 361 (2014) 68–73. https://doi.org/10.1016/j.jmmm.2014.02.075

[255]   P. Zhang, S. Hua, H. Yang, M. Yue and H. Xu, Effect of Ce doping on the
magnetic and magnetocaloric properties of $Pr_{0.5}Sr_{0.5-x}Ce_xMnO_3$ manganites,
Phase Transitions, 87 (2014) 357–362.
https://doi.org/10.1080/01411594.2013.815355

[256]   G. Tonozlis and G. Litsardakis, The structural, magnetic and magnetocaloric
properties of Ba doped La manganites, Physica Status Solidi C, 11 (2014) 1133–
1138. https://doi.org/10.1002/pssc.201300720

[257]   A. Mleiki, S. Othmani, W. Cheikhrouhou-Koubaa, M. Koubaa, A.
Cheikhrouhou, E.K.Hlil, Effect of praseodymium doping on the structural,
magnetic and magnetocaloric properties of $Sm_{0.55}Sr_{0.45}MnO_3$ manganite, Solid
State Communications, 223 (2015) 6–11.
https://doi.org/10.1016/j.ssc.2015.08.019

[258]   A. Jerbi, A. Krichene, N. Chniba-Boudjada, W. Boujelben, Magnetic and
magnetocaloric study of manganite compounds $Pr_{0.5}A_{0.05}Sr_{0.45}MnO_3$ (A=Na and
K) and composite, Physica B, 477 (2015) 75–82.
https://doi.org/10.1016/j.physb.2015.08.022

[259]   A. Mleiki, S. Othmani, W. Cheikhrouhou-Koubaa, M. Koubaa, A.
Cheikhrouhou, E.K.Hlil, Effect of praseodymium doping on the structural,
magnetic and magnetocaloric properties of $Sm_{0.55-x}Pr_xSr_{0.45}MnO_3$ ($0.1 \leq x \leq 0.4$)
manganites, Journal of Alloys and Compounds, 645 (2015) 559–565.
https://doi.org/10.1016/j.jallcom.2015.05.043

[260]   S. Vadnala, P. Pal, S. Asthana, Influence of A-site cation disorder on structural and magnetocaloric properties of $Nd_{0.7-x}La_xSr_{0.3}MnO_3$ (x = 0.0, 0.1, 0.2 & 0.3), Journal of Rare Earths, 33 (2015) 1072. https://doi.org/10.1016/S1002-0721(14)60528-7

[261]   A. Mehri, W. C. Koubaa, M. Koubaa1, A. Cheikhrouhou, Magnetocaloric Properties in $La_{0.5}Ca_{0.45}K_{0.05}MnO_3$, $Pr_{0.5}Sr_{0.45}K_{0.05}MnO_3$, and $Nd_{0.5}Sr_{0.45}K_{0.05}MnO_3$ Manganites, Journal of Superconductivity and Novel Magnetism, 28 (2015) 3135– 139. https://doi.org/10.1007/s10948-015-3138-y

[262]   N. Nedelko, S. Lewinska, A. Pashchenko, I. Radelytskyi, R. Diduszko, E. Zubov, W. Lisowski, J. W. Sobczak, K. Dyakonov, A. S´lawska-Waniewska, V. Dyakonov, H. Szymczak, Magnetic properties and magnetocaloric effect in $La_{0.7}Sr_{0.3-x}Bi_xMnO_3$ manganites, Journal of Alloys and Compounds, 640 (2015) 433–439. https://doi.org/10.1016/j.jallcom.2015.03.126

[263]   A. Varvescu, I. G. Deac, Critical magnetic behavior and large magnetocaloric effect in $Pr_{0.67}Ba_{0.33}MnO_3$ perovskite manganite, Physica B, 470-471 (2015) 96–101. https://doi.org/10.1016/j.physb.2015.04.037

[264]   S. C. Maatar, R. M'nassri, W. C. Koubaa, M. Koubaa, A. Cheikhrouhou, Structural, magnetic and magnetocaloric properties of $La_{0.8}Ca_{0.2-x}Na_xMnO_3$ manganites ($0 \leq x \leq 0.2$), Journal of Solid State Chemistry, 225 (2015) 83–88. https://doi.org/10.1016/j.jssc.2014.12.007

[265]   K. Sbissi, M. L. Kahn, M. Ellouze, E. K. Hlil, F. Elhalouani, The Magnetic and magnetocaloric properties of $Pr_{1-x}Bi_xMnO_3$ (x = 0.2 and 0.4) manganites, Journal of Superconductivity and Novel Magnetism, 28 (2015) 1433–1438. https://doi.org/10.1007/s10948-015-2985-x

[266]   Z. U. Rehman, M. S. Anwar, B. H. Koo, Influence of barium doping on the magnetic and magnetocaloric properties of $Pr_{1-x}Ba_xMnO_3$, Journal of Superconductivity and Novel Magnetism, 28 (2015)1629–1634. https://doi.org/10.1007/s10948-014-2933-1

[267]   T. Raoufi, M. H. Ehsani, D. S. Khoshnoud, Magnetocaloric properties of $La_{0.6}Sr_{0.4}MnO_3$ prepared by solid state reaction method, Journal of Alloys and Compounds, 689 (2016) 65 – 873. https://doi.org/10.1016/j.jallcom.2016.08.063

[268]   H. Omrani, M. Mansouri, W. Cheikhrouhou Koubaa, M. Koubaa, A. Cheikhrouhou, Structural, magnetic and magnetocaloric investigations in $Pr_{0.6}$-

$_x$Er$_x$Ca$_{0.1}$Sr$_{0.3}$MnO$_3$ ($0 \leq x \leq 0.06$) manganites, Journal of Alloys and Compounds, 688 (2016) 752– 761. https://doi.org/10.1016/j.jallcom.2016.07.082

[269] G. Akça, S. K. Çetin, M. Güneş, A. Ekicibil, Magnetocaloric properties of (La$_{1-x}$Pr$_x$)$_{0.85}$K$_{0.15}$MnO$_3$ (x = 0.0, 0.1, 0.3 and 0.5) perovskite manganites, Ceramics International,42 (2016) 19097–19104. https://doi.org/10.1016/j.ceramint.2016.09.070

[270] A. Elghoul, A. Krichene, W. Boujelben, Electrical, magnetic and magnetocaloric properties of polycrystalline Pr$_{0.63}$A$_{0.07}$Sr$_{0.3}$MnO$_3$ (A = Pr, Sm and Bi), Journal of Physics and Chemistry of Solids, 98 (2016) 263–270. https://doi.org/10.1016/j.jpcs.2016.07.021

[271] H. B. Khlifa, S. Othmani, I. Chaaba, S. Tarhouni, W. Cheikhrouhou-Koubaa, M. Koubaa, A. Cheikhrouhou, E. K. Hlil, Effect of K-doping on the structural, magnetic and magnetocaloric properties of Pr$_{0.8}$Na$_{0.2-x}$K$_x$MnO$_3$ ($0 \leq x \leq 0.15$) manganites, Journal of Alloys and Compounds, 680 (2016) 388 – 396. https://doi.org/10.1016/j.jallcom.2016.04.138

[272] P. T. Phong, L. T. Duy, L. V. Bau, N. V. Dang, D. H. Manh, In-Ja Lee, Magnetic and magnetocaloric properties of selected Pb-doped manganites, Journal of Electroceramics, 36 (2016) 58– 64. https://doi.org/10.1007/s10832-016-0026-1

[273] S. K. Çetin, M. Acet, A. Ekicibil, Effect of Pr-substitution on the structural, magnetic and magnetocaloric properties of (La$_{1-x}$Pr$_x$)$_{0.67}$Pb$_{0.33}$MnO$_3$ ($0.0 \leq x \leq 0.3$) manganites, Journal of Alloys and Compounds, 727 (2017) 1253–1262. https://doi.org/10.1016/j.jallcom.2017.08.199

[274] G. Akça, S. K. Çetin, A. Ekicibil, Structural, magnetic and magnetocaloric properties of (La$_{1-x}$Sm$_x$)$_{0.85}$K$_{0.15}$MnO$_3$ (x = 0.0, 0.1, 0.2 and 0.3) perovskite manganites, Ceramics International, 43 (2017) 15811–15820. https://doi.org/10.1016/j.ceramint.2017.08.150

[275] I. S. Debbebi1, W. C. Koubaa, A. Cheikhrouhou, E. K. Hlil, Structural, magnetic and magnetocaloric investigation of La$_{0.7-x}$Dy$_x$Ca$_{0.3}$MnO$_3$ (x = 0.00, 0.01 and 0.03) manganite, Journal of Material Science: Material in Electronics, 28 (2017) 16965–16972. https://doi.org/10.1007/s10854-017-7618-7

[276] A. Modi, N. K. Gaur, Effect of Sm substitution on magnetic and magnetocaloric properties of La$_{0.7-x}$Sm$_x$Ba$_{0.3}$MnO$_3$ ($0 \leq x \leq 0.2$) compounds, Journal of Magnetism and Magnetic Materials, 441 (2017) 217–221. https://doi.org/10.1016/j.jmmm.2017.05.073

[277]   T.A. Ho, S. H. Lim, C. M. Kim, M. H. Jung, T. O. Ho, P. T. Tho, T. L. Phan, S. C. Yu, Magnetic and magnetocaloric properties of $La_{0.6}Ca_{0.4-x}Ce_xMnO_3$, Journal of Magnetism and Magnetic Materials, 438 (2017) 52–59. https://doi.org/10.1016/j.jmmm.2017.04.038

[278]   R. Thaljaoui, M. Pękała, J.-F. Fagnard, Ph. Vanderbemden, Effect of Ag substitution on structural, magnetic and magnetocaloric properties of $Pr_{0.6}Sr_{0.4-x}Ag_xMnO_3$ manganites, Journal of Rare Earths, 35 (2017) 875. https://doi.org/10.1016/S1002-0721(17)60989-X

[279]   L. Han, F. Liang, H. Zhu, H. Pang, J. Yang, T. Zhang and Z. Yan, Structural, magnetic and magnetocaloric properties of electron-doped manganite $La_{0.9}Hf_{0.1}MnO_3$, Material Research Express, 4 (2017) 086103. https://doi.org/10.1088/2053-1591/aa806a

[280]   M. Zarifi, P. Kameli, M. Mansouri, H. Ahmadvand, H. Salamati, Magnetocaloric effect and critical behavior in $La_{0.8-x}Pr_xSr_{0.2}MnO_3$ ($x = 0,2, 0,4, 0,5$) manganites, Solid State Communications, 262 (2017) 20–28. https://doi.org/10.1016/j.ssc.2017.06.007

[281]   C.G. Ünlü, Y. E. Tanıs, M. B. Kaynar, T. Simsek, S. Ozcan, Magnetocaloric effect in $La_{0.7}Nd_xBa_{(0.3-x)}MnO_3$ ($x = 0.00, 0.05, 0.1$) perovskite manganites, Journal of Alloys and Compounds, 704 (2017) 58– 63. https://doi.org/10.1016/j.jallcom.2017.02.030

[282]   I. Sfifir, A. Ezaami, W. C. Koubaa, A. Cheikhrouhou, Structural, magnetic and magnetocaloric properties in $La_{0.7-x}Dy_xSr_{0.3}MnO_3$ manganites ($x = 0.00, 0.01$ and 0.03), Journal of Alloys and Compounds, 696 (2017) 760–767. https://doi.org/10.1016/j.jallcom.2016.11.286

[283]   S. Tarhouni, A. Mleiki, I. Chaaba, H. Ben Khelifa, W. Cheikhrouhou-Koubaaa, M. Koubaaa, A. Cheikhrouhoua, E. K. Hlil, Structural, magnetic and magnetocaloric properties of Ag-doped $Pr0.5Sr_{0.5-x}Ag_xMnO3$ manganites ($0.0 \leq x \leq 0.4$), Ceramics International, 43 (2017) 133–143. https://doi.org/10.1016/j.ceramint.2016.09.122

[284] L. Han, S. Pang, H. Zhu, P. Zhang, J. Yang, T. Zhang, Magnetocaloric effect and critical properties in $La_{0.85}Li_{0.15}MnO_3$, Journal of Materials Science: Materials in Electronics, 29 (2018) 20156–20161. https://doi.org/10.1007/s10854-018-0148-0

[285] P. D. H. Yena, N. T. Dung, T. D. Thanh, S-C. Yu, Magnetic properties and magnetocaloric effect of Sr-doped $Pr_{0.7}Ca_{0.3}MnO_3$ compounds, Current Applied Physics, 18 (2018) 1280–1288. https://doi.org/10.1016/j.cap.2018.07.006

[286] I. Walha, M. Smari, T. Mnasri, E. Dhahri, Structural, magnetic, and magnetocaloric properties of Ag-doped in the $La_{0.6}Ca_{0.4}MnO_3$ compound, Journal of Magnetism and Magnetic Materials, 454 (2018) 190–195. https://doi.org/10.1016/j.jmmm.2018.01.084

[287] K. Snini, F. B. Jemaa, M. Ellouze, E.K. Hlil, Structural, magnetic and magnetocaloric investigations in $Pr_{0.67}Ba_{0.22}Sr_{0.11}Mn_{1-x}Fe_xO_3$ ($0 \leq x \leq 0.15$) manganite oxide, Journal of Alloys and Compounds, 739 (2018) 948-954. https://doi.org/10.1016/j.jallcom.2017.12.309

[288] S. Vadnala, S. Asthana, Magnetocaloric effect and critical field analysis in Eu substituted $La_{0.7-x}Eu_xSr_{0.3}MnO_3$ (x = 0.0, 0.1, 0.2, 0.3) manganites, Journal of Magnetism and Magnetic Materials, 446 (2018) 68–79. https://doi.org/10.1016/j.jmmm.2017.09.001

[289] V. Franco, J.S. Blázquez, J.J. Ipus, J.Y. Law, L.M. Moreno-Ramírez, A. Conde, Magnetocaloric effect: From materials research to refrigeration devices, Progress in Materials Science, 93 (2018) 112–232. https://doi.org/10.1016/j.pmatsci.2017.10.005

Keyword Index

Abut the Authors

Prof. Dr. Hüseyin Gencer was born in a rural town of Malatya (Turkey) in 1968. He completed his basic education in his town. He graduated from İnönü University, Faculty of Arts and Sciences, Department of Physics (Malatya, Turkey) with high success. He completed his master's degree at Inonu University, Institute of Science in 1992. He received his PhD degree at Inonu University Science Institute in 1998. He has nearly 30 years of academic experience in physics and has been working as a physics professor in İnönü University, Science and Arts Faculty, Department of Physics since 2013. His interests are generally magnetic materials, magnetocaloric effect, magnetocaloric materials, magnetoresistance. He has done many studies on the structural, magnetic, transport, magnetocaloric and magnetoresistance properties of many different ferromagnetic materials. In particular, he specializes in the structural, magnetic, transport and magnetocaloric properties of perovskite manganites. He published 36 articles, mostly in peer-reviewed International Journals.

Prof. Dr. Veli Serkan Kolat was born in Adana, Turkey in 1975. He graduated from Inonu University (Malatya, Turkey) Physics Department in 1998 and completed his master's degree at Malatya Inonu University Science Institute in 2002. He received his PhD degree at Inonu University Science Institute in 2007. He is currently a Professor in Inonu University, Science Faculty and Department of Physics, Malatya, Turkey. His research interests are in the area of magnetic materials, magnetocaloric effect and magnetic sensors. He has done many studies on the magnetic and structural properties of ferromagnetic materials. To his credit, he published 51 research papers in peer-reviewed International Journals.

Prof. Dr. Tekin Izgi is a professor of Physics, Department of Physics, University School of Sciences at Inonu University, Malatya, Turkey. He obtained his B.Sc., from Inonu University, Malatya and Ph.D. from Anadolu University, Eskişehir, Turkey. His research interests are on magnetic properties of ferromagnetic materials, spectroscopic and quantum mechanical calculations of atomic and molecular structures, and condensed matter physics. To his credit he published 46 research articles in various International Journals. Most of these publications are on investigation of the magnetic and structural properties of ferromagnetic materials.

Prof. Dr. Nevzat Bayri graduated from Inonu University (Malatya, Turkey) Physics Department in 1998 and completed his master's degree at Malatya Inonu University Science Institute in 2002. He obtained a PhD degree in 2007 on his thesis entitled "Stress-Impedance and Magnetoimpedance Effect in Amorphous Ferromagnetic Alloys". He is currently working as a professor in the Department of Mathematics and Science Education, Inonu University, Turkey. He published 35 research papers in diffrent International scientific Journals. His research work includes amorphous ferromagnetic alloys, magnetic sensors, magnetoimpedance and magnetocaloric effect.

Dr. Selcuk Atalay was born in Turkey in 1964. He got his PhD in magnetoelastic properties of amorphous wires in 1992 from Bath University, United Kingdom. His interests are oriented to magnetocaloric materials, amorphous ferromagnetic wires and ribbons. His research interests also focus on magnetic measurements and magnetic sensors, especially magnetoimpedance and magnetoelastic effects. He has (co-)authored over 100 peer-reviewed papers in magnetism. At present, he is working as a lecturer and also he is head of applied magnetism group at Physics Department, Inonu University, Turkey.

Selçuk atalay received the young scientist award from The Scientific and technological Research Council of Turkey, TUBITAK, in 2000 and the young researcher incentive award from Turkey Science Acedemy , TUBA, in 2002 for outstanding contributions in the field of magnetic materials and application. In addition, he received fellowships from Royal Society,U.K. and TUBITAK.

About the Editors

Dr. Rajshree B. Jotania is a professor of Physics, Department of Physics, Electronics and Space science, University School of Sciences at Gujarat University, Ahmedabad, India. She obtained her B.Sc., M.Sc. and Ph.D from Saurashtra University, Rajkot, India. She was Junior Research Fellow (DAE-BRNS project) during 1987 to 1989 at Physics Department, Saurashtra University, Rajkot, India. She obtained a few regional, and national awards for contribution toward scientific research. She worked at National Chemical Laboratory, Pune, India for two months as a Summer Visiting Teacher Fellow in 2005 and as a Visiting Scientist fellow in 2011. She possesses 30 years of teaching experience at UG and PG level. She is a member of Board of studies at few Universities of Gujarat, India and a Mentor of DST-INSPIRE (Department of Science and Technology- Innovation in Science Pursuit for Inspired Research) program. She has published more than 100 papers in various research journals and conference proceedings. She has delivered more than 20 invited talks at various DST-INSPIRE Internship science camp in India. She has edited three books entitled 'Ferrites and ceramic composites' (Vol. I & II, Trans Tech Publisher, Switzerland) and Magnetic Oxides and Ceramic Composites (MRF, USA). She has visited Singapore, Malaysia, New York and North Africa for research work. She has attended more than 50 international, national conferences/symposiums/seminars/ academy meeting and worked as a chair person as well as delivered invited talks in a few international and national conferences. She possesses a life membership of eight professional bodies and she has guided six Ph. D, ten M. Phil students. At present few more students are working under her guidance for M.Phil and Ph. D. To date she has completed five research projects of various agencies. She has worked as deputy co-coordinator, DRS (SAP-I) program. She is an active member of Indian Association of Physics Teacher.

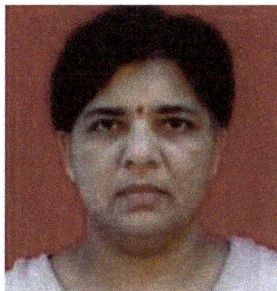

S. H. Mahmood obtained his B.Sc. degree in Physics from The University of Jordan, Amman in 1978, and his PhD degree in Physics from Michigan State University, East Lansing, Michigan, USA in 1986. Between 1986 and 2010, he was a faculty member at Yarmouk University, Irbid, Jordan, and since 2010, he is a professor of physics at The University of Jordan, Amman. During his academic career, he was involved in teaching, research, graduate work supervision, and administration. He held the positions of Director of the Center for Theoretical and Applied Physical Sciences, Chairman of Physics, Dean of Science, Dean of Scientific research and Graduate Studies, and Vice President. He published more than 130 articles in peer-reviewed international journals, participated in tens of regional and international conferences, and supervised tens of M.Sc. and PhD theses. He also received several national, regional and international Awards and Honors for Academic excellence and contribution to science. Also, he participated in the management and execution of nationally and internationally funded projects concerned with establishing long-term research programs, new academic programs, capacity building, and curricular development. Additionally, he actively participated as a scientific advisor, and a member of scientific committees and councils of Scientific Research Funds in Jordan, and of editorial boards of international journals.

www.ingramcontent.com/pod-product-compliance
Lightning Source LLC
Chambersburg PA
CBHW071716210326
41597CB00017B/2501